TERRORISM, HOT SPOTS AND C(

ARMED CONFLICT IN THE 21ST CENTURY

THE INFORMATION REVOLUTION AND POST-MODERN WARFARE

TERRORISM, HOT SPOTS AND CONFLICT-RELATED ISSUES

Additional books in this series can be found on Nova's website under the Series tab.

Additional e-books in this series can be found on Nova's website under the e-books tab.

Terrorism, Hot Spots and Conflict-Related Issues

Armed Conflict in the 21st Century

The Information Revolution and Post-Modern Warfare

Steven Metz

Copyright © 2018 by Nova Science Publishers, Inc.

All rights reserved. No part of this book may be reproduced, stored in a retrieval system or transmitted in any form or by any means: electronic, electrostatic, magnetic, tape, mechanical photocopying, recording or otherwise without the written permission of the Publisher.

We have partnered with Copyright Clearance Center to make it easy for you to obtain permissions to reuse content from this publication. Simply navigate to this publication's page on Nova's website and locate the "Get Permission" button below the title description. This button is linked directly to the title's permission page on copyright.com. Alternatively, you can visit copyright.com and search by title, ISBN, or ISSN.

For further questions about using the service on copyright.com, please contact:
Copyright Clearance Center
Phone: +1-(978) 750-8400   Fax: +1-(978) 750-4470   E-mail: info@copyright.com.

NOTICE TO THE READER

The Publisher has taken reasonable care in the preparation of this book, but makes no expressed or implied warranty of any kind and assumes no responsibility for any errors or omissions. No liability is assumed for incidental or consequential damages in connection with or arising out of information contained in this book. The Publisher shall not be liable for any special, consequential, or exemplary damages resulting, in whole or in part, from the readers' use of, or reliance upon, this material. Any parts of this book based on government reports are so indicated and copyright is claimed for those parts to the extent applicable to compilations of such works.

Independent verification should be sought for any data, advice or recommendations contained in this book. In addition, no responsibility is assumed by the publisher for any injury and/or damage to persons or property arising from any methods, products, instructions, ideas or otherwise contained in this publication.

This publication is designed to provide accurate and authoritative information with regard to the subject matter covered herein. It is sold with the clear understanding that the Publisher is not engaged in rendering legal or any other professional services. If legal or any other expert assistance is required, the services of a competent person should be sought. FROM A DECLARATION OF PARTICIPANTS JOINTLY ADOPTED BY A COMMITTEE OF THE AMERICAN BAR ASSOCIATION AND A COMMITTEE OF PUBLISHERS.

Additional color graphics may be available in the e-book version of this book.

Library of Congress Cataloging-in-Publication Data

ISBN: 978-1-53613-703-3

Published by Nova Science Publishers, Inc. † New York

Contents

Foreword   vii
Douglas C. Lovelace, Jr.

Summary   ix

Introduction   xxiii

Part I: Strategic Context   1

Part II: Images of Future War   21

Part III: The Mark of Success for Future Militaries   63

Part IV: Conclusion and Recommendations   81

Endnotes   85

About the Author   103

Index   105

FOREWORD*

Within the past decade, the U.S. military has implemented a number of programs to assess the changes underway in the global security environment and in the nature of warfare. Defense leaders and thinkers have concluded that revolutionary change is taking place and, if the United States develops appropriate technology, warfighting concepts, and military organizations, it can master or control this change, thus augmenting American security.

In this monograph, Dr. Steven Metz, who was one of the earliest analysts of the strategic dimension of the revolution in military affairs, suggests that official thinking within the U.S. military may be too narrow. The information revolution, he contends, will have far-reaching strategic effects. The transformation it brings will not only be technological, but political, social, ethical and strategic as well.

As he explores the impact that the information revolution may have on the conduct of armed conflict, Dr. Metz introduces a number of ideas which need further analysis, including the potential for the emergence of nontraditional, networked enemies; multidimensional asymmetry; the privatization of security; and the potential impact of technologies like robotics, nonlethality, and nanotechnology. He concludes with an

---

* This is an edited, reformatted and augmented version of "Armed Conflict in the 21st Century - The Information Revolution and Post-Modern Warfare" by Steven Metz, originally published by the Strategic Studies Institute in 2000.

assessment of the features likely to characterize successful militaries in the 21st century.

Because it deals with the future, this study is conceptual and speculative. But the issues and linkages it raises are directly relevant to today's strategic thinkers and leaders. The Strategic Studies Institute is pleased to offer it as a contribution to debate over the nature of the challenges that the U.S. military will face in coming decades.

Douglas C. Lovelace, Jr.
Interim Director
Strategic Studies Institute

SUMMARY

INTRODUCTION

The information revolution is increasing interconnectedness and escalating the pace of change in nearly every dimension of life. By examining the ongoing changes in the nature of armed conflict and thinking expansively, looking for wider implications and relationships, and exploring cross-cutting connections between technology, ethics, social trends, politics, and strategy, the architects of the future U.S. military can increase the chances of ultimate success.

PART I: STRATEGIC CONTEXT

Interconnectedness and Globalization

One of the most important changes associated with the information revolution is a dramatic increase in the interconnectedness of people around the world. Almost no dimension of modern life has been untouched by the information revolution. In the realm of security, the information revolution brings both good news and bad news, speeding the accumulation of information and slowing the pace of decision-making.

The information revolution has also sped up the pace of change in all aspects of life. Rapid change always has winners and losers. Much of the violence that will exist in the early 21st century will originate from the losers of the change underway today.

Organizational Change

The information revolution is altering the shape of economic and political organizations. Today, the successful commercial firm is one with a global perspective, a web of strategic partnerships, and internal flexibility based on project teams or work groups rather than hierarchies or bureaucracies. This phenomenon is migrating to the political world as well. Clinging to old practices and organizations entails escalating costs and risks for governments as much as for corporations. At the same time that interconnectedness undercuts the viability of authoritarianism by allowing repressed citizens to communicate, organize, and mobilize, it also places handcuffs on elected governments. This reflects an historic deconcentration of political, economic, and ethical power.

The information revolution is both a force for stability and for instability. On the positive side, it complicates the task of old-style repression and facilitates the development of grass roots civil society. But the information revolution also allows organizations intent on instability or violence to form alliances, thus making the world more dangerous. Some of the most complex struggles of the 21st century will pit polyglot networks against states. Hierarchies and bureaucracies face serious disadvantages when pitted against unscrupulous, flexible, adaptable enemies.

The Changing Nature of Armed Conflict

The U.S. Department of Defense and the military services hold that speed, knowledge, and precision will minimize casualties and lead to the rapid resolution of wars, thus minimizing the problems associated with the challenges to the political utility of force. States with fewer intellectual and financial resources than the United States will not have the luxury of using technology to solve strategic problems. Whether the United States can be deterred from intervention by weapons of mass destruction or terrorism is one of the central questions for the future global security environment.

Privatization

Interconnectedness, the dispersion of power and knowledge that flows from the information revolution, and the eroding legitimacy of armed force are leading toward a multidimensional trend toward privatization within the realms of security and armed conflict. As nations seek ways to attain a surge capacity without the expense of sustaining a large, peacetime military, and as they face difficulties recruiting from their own populations, contracting will be an attractive option for filling the ranks. Corporate armies, navies, air forces, and intelligence services may be major actors in 21st century armed conflict. This will open new realms of strategy and policy.

Asymmetry

States which decide to commit aggression in coming decades will know that if the United States and the world community decide to counter the aggression, they can. The qualitative gap between the U.S. military and all others is wide and growing. This leaves aggressors two options: they can pursue indirect or camouflaged aggression, or they can attempt to deter or counter American intervention asymmetrically. Asymmetry is a characteristic of periods of rapid change, particularly revolutionary ones. In geological history, there have been times when many new species emerged. Most proved unable to survive, leading to new periods with less diversity. Military history follows the same pattern: periods of great diversity follow periods of relative homogeneity. The current era is one of diversity. For the period of diversity, asymmetry will be a dominant characteristic of armed conflict.

Combatants

In the opening half of the 21st century, the types of state and nonstate combatants which have characterized recent armed conflict will continue to exist, but they are likely to be joined by new forms. The U.S. military probably will be the first *post-modern state* combatant, attaining greatly amplified speed and precision by the integration of information technology and development of a system of systems which links together methods for target acquisition, strikes, maneuver, planning, communication, and supply. Its organization will be less rigidly hierarchical than that of modern state combatants. The final type of combatants in 21st century armed conflict are likely to be *post-modern nonstate* ones. These will consist of loose networks of a range of nonstate organizations, some political or ideological in orientation, others seeking profit.

PART II: IMAGES OF FUTURE WAR

The Official View

Broadly speaking, the opening decades of the 21st century are likely to see some combination of three modes of warfare: formal war, informal war, and gray area war. Formal war pits state militaries against other state militaries. It has been the focus of most futures-oriented thinking within the U.S. military and Department of Defense. The official vision of future war reflects the belief that "information superiority" will be the lifeblood of a post-modern military and thus the key to battlefield success.

Futures-oriented thinking deals with force development which is a responsibility of the services. In fact, most of the futures thinking within the U.S. military is still done by the services. So far, the Army's program is the most elaborate. It has formulated a vision that is highly innovative in its approach to technology, organization, and leadership, but conservative in its assumptions about the nature of warfare and the purposes of American military power. The U.S. Air Force's vision of future war is also characterized by a combination of creativity and conservatism. The Marines are looking at fairly radical changes in tactical and operational procedures, including new organizations and doctrine. The Navy's view of future war is based on a "revolution in strike warfare" using existing major platforms with better systems of target acquisition, intelligence, and guidance.

The official American view of the future consistently treats technology, particularly information technology, as a force multiplier rather than as a locomotive for revolutionary transformation. With the exception of adding three new tasks for the U.S. military—space operations, information warfare, and homeland protection—the official vision anticipates few if any strategic shifts.

Asymmetry Again

Asymmetry has become a central concept in official American thinking about future warfare. The question then becomes: what forms of asymmetry will be most common and, more importantly, most problematic for the United States? Enemies using precision munitions or weapons of mass destruction to complicate deployment into a theater of operations could pose a serious challenge to some of the most basic tenets of American strategy. A counter-deployment strategy is only one of several asymmetric approaches that future enemies may attempt. They might also resort to terrorism, either in conjunction with a counter deployment strategy or in lieu of it. Of all forms of asymmetry, urban warfare may be the most problematic and the most likely. Two types of technology, though, might help alleviate some of the challenges posed by urban operations: nonlethal weapons and robotics.

Broadly speaking, the opening decades of the 21st century will see both symmetric formal war pitting two modern states, and asymmetric formal war pitting a post-modern military against a modern one. It remains to be seen whether another post-modern military will emerge to challenge the United States or whether, as American strategic thinking posits, the post-modern U.S. military will always be able to overcome the asymmetric methods used by modern militaries.

Informal War

Informal war is armed conflict where at least one of the antagonists is a nonstate entity such as an insurgent army or ethnic militia. Twenty-first century informal war will be based on some combination of ethnicity, race, regionalism, economics, personality, and ideology. Informal war will both be more common and more strategically significant. Combat in future informal war is likely to remain "hands on," pitting the combatants in close combat. Warriors will be interspersed among noncombatants, using them as shields and bargaining chips. At times, refugee disasters will be deliberately stoked and sustained to attract outside attention and intervention. Unlike formal war, informal war will remain dirty and bloody.

Some types of informal war will be comparatively simple. Counterinsurgency, which uses military forces to attain not only the short-term restoration of order but also ultimate resolution of the conflict that led to disorder in the first place, is a different and more difficult matter. There is no doctrine or strategy for dealing with networked opponents, be they existing criminal cartels or future insurgents. To be successful against future insurgents, the U.S. military will need better intelligence, better force protection, and greater precision at the tactical and strategic levels. In part, these things require new organizational methods. Emerging technology also holds promise. Again, nonlethal weapons and robotics may prove the most vital.

Gray Area War

Gray area phenomena combine elements of traditional warfighting with those of organized crime. Today, gray area threats are increasing in strategic significance. Since gray area war overlaps and falls in between traditional national security threats and law enforcement issues, states must often scramble to find the appropriate security structure to counter it. As the debate within the United States over the use of the military to counter gray area enemies intensifies in coming years, creation of an American national gendarmerie should be considered. It could form its own alliances with similar security forces around the world and operate more effectively against gray area enemies in an interconnected security environment and globalized economy.

Strategic Information Warfare

Future war may see attacks via computer viruses, worms, logic bombs, and Trojan horses rather than bullets, bombs, and missiles. Information technology might provide a politically usable way to damage an enemy's national or commercial infrastructure badly enough to attain victory without having to first defeat fielded military forces.

Today strategic information warfare remains simply a concept or theory. The technology to wage it does not exist. But until it is proven ineffective, states and nonstate actors which have the capacity to attempt it probably will, doing so because it appears potentially effective and less risky than other forms of armed conflict.

Cyberattacks might erode the traditional advantage large and rich states hold in armed conflict. Private entities might be able to match state armed forces. As one of the world's most "wired" nations, strategic information warfare could be particularly problematic for the United States, forcing policymakers and military strategists to examine some of their most basic beliefs about warfighting and national security.

Technological Transformation

New technologies or new combinations of technology have the potential to alter not only tactics and operational methods, but military strategy itself. Soon technology may allow military planners to select which individual or physical object in a building is to be destroyed.

Coming decades are likely to see the proliferation of robots around the world and in many walks of life. As one of the most avid customers of new technology, this will certainly affect the American military. Initially, the prime function of military robots will be to replace humans in particularly dangerous or tedious functions. The real breakthrough and decision point will come when robots advance to the point that they have the potential for combat use.

While initial thinking about robotics concentrates on miniaturization and the integration of networks of small robots with relatively limited functions, partially organic robots may prove nearly as useful. Beyond technological obstacles, the potential for effective battlefield robots raises a whole series of strategic, operations, and ethical issues, particularly when or if robots change from being lifters to killers.

Other emerging technologies could prove equally revolutionary. One example is what can be called "psychotechnology." Future military commanders might have the technology to alter the beliefs, perceptions, and feelings of enemies. Any developments in this realm warrant very close scrutiny. Barring some sort of truly fundamental change in the global security environment, they should be eschewed.

PART III: THE MARK OF SUCCESS FOR FUTURE MILITARIES

Foundation

Even in revolutionary times, continuity outweighs change. This holds true for the current revolution in military affairs. War will always involve a dangerous and dynamic relationship among passion, hatred, reason, chance and probability. The "specialness" of warfighting and warriors will survive any real or apparent changes in the nature of armed conflict. War is and will be distinct from other types of human activity. Largely because of this, future warriors, at least in democracies like the United States, will continue to be bound by an ethos stressing duty, honor, sacrifice, and the highest ethical standards.

Speed

One of the most important determinants of success for 21st century militaries will be the extent to which they are faster than their opponents. Tactical and operational speed comes from information technology—the "digitized" force—and appropriate doctrine and training. *Strategic* speed will be equally important as a determinant of success in future armed conflict. For nations that undertake long-range power projection, strategic speed includes mobility into and within a theater of military operations. Strategic speed also entails faster decision-making. One of the core dilemmas the United States is likely to face is having a military that can deploy and operate at lightning speed, while strategic and political decision-making remains a time-consuming process of consensus building.

Speed also has an even broader, "meta-strategic" meaning. The militaries which meet with the greatest success in future armed conflict will be those which can undertake rapid organizational and conceptual adaptation. Successful state militaries must institutionalize procedures for what might be called "strategic entrepreneurship"—the ability to rapidly identify and understand significant changes in the strategic environment and form appropriate organizations and concepts.

Precision

So far thinking on the revolution in military affairs has focused on what might be called *physical* precision—the ability to hit targets with great accuracy from great distances with precisely the desired physical effect. Military strategists and commanders must come to think in terms of *psychological* precision as well: shaping a military operation so as to attain the desired attitudes, beliefs, and perceptions on the part of both the enemy and other observers, whether noncombatants in the area of operations or global audiences.

Precision has a strategic component which is sometimes overlooked. Strategic precision entails shaping a military so that it best reflects its nation's strategic situation, including strategic culture, level of technological development, and most significant threats. For the U.S. military, this entails finding the appropriate balance among capabilities to deal with formal war, informal war, and gray area war. It also entails reaching a degree of privatization which maximizes efficiency without creating unacceptable risks. In attaining strategic precision, past success can be a hindrance. Victory limits the urge to adapt and innovate. For the United States, avoiding a victory-induced slumber will be a key step toward a post-modern military.

Finding and Hiding

One of the most crucial dynamics of future armed conflict will be the struggle between finding and hiding. Successful militaries will be those better at finding their enemies than their enemies are at finding them. Within the United States, the emphasis has been on the offensive part of this equation—the finding. Hiding, though, warrants more attention. Future military strategists must rebuild their understanding of deception and hiding, working with new information technology that allows things like morphing and sophisticated spoofing (including things like holographic soldiers, tanks, planes, and so forth). In particular, the notions of operational and strategic deception must be revisited.

Reorganizing

The most successful future militaries will be those that undertake a "blank slate" reevaluation of their most basic concepts and organizational precepts. Developing hybrid blends of hierarchical structures with networks, public components with private, and humans with machines will be particularly important. Reevaluating career paths in the military also might be necessary. The trend in the commercial world has been toward a blurring between management and staff. If this is extrapolated to the military, it might be necessary to consider whether the division of a service into enlisted personnel and commissioned officers makes sense in the 21st century. In addition, the organization of militaries into land, sea, and air services needs assessed. Perhaps it would make more sense to organize them into components focused on a specific type of armed conflict—one for formal war, one for informal, and one for gray area war. Alternatively, post-modern militaries must consider whether a new service is needed for new operating environments. Those militaries able to let go of old organizational patterns and embrace, even master new ones will be the most likely to succeed in future armed conflict.

Adjusting Civil-Military Relations

The current health of American civil-military relations is based on the perception that: (1) the military has a vital job to do in defending the nation against external enemies, it does so very competently, and should receive adequate resources to do so; (2) the effectiveness of the U.S. military does not threaten domestic civil rights or political institutions; and, (3) the U.S. military represents the best of traditional American values. Changes in any of these three components could degrade civil-military relations.

The U.S. military must do its part to help forestall problems with civil-military relations. Foremost, it must assure that any capabilities or methods it develops reflect national values and strategic culture. For instance, it should eschew operational concepts that call for the preemptive use of force on the part of the United States or for actions that would indiscriminately harm noncombatants. And, unless circumstances change in some fundamental way, the military should eschew development of dangerous new technologies like psychotechnology which run counter to American values like personal privacy and civil rights.

Controlling for Asymmetry

Since asymmetric conflict will be common in the opening decades of the 21st century, finding ways to resist or transcend it will be one of the determinants of success for militaries and other organizations that participate in armed conflict. For the United States, what might be called an asymmetry of time is likely to be particularly problematic. Today, long wars are simply considered inconceivable in American planning While everything suggests that the future United States (just like the current one) would prefer short wars, failing to plan for protracted conflict increases the chances it will occur. Given this, greater attention should be given to protracted war in the various wargames, seminars, and simulations that the U.S. military uses to think about future armed conflict.

Adapting to Technological Shifts

The ability to accept and capitalize on emerging technology will be a determinant of success in future armed conflict. No military is better at this than the American, in large part because no culture is better at it than the American. That said, there will be new, radical technologies with great promise which will challenge the ability of the military to master and integrate. In particular, robotics, miniaturization, and nonlethality are likely to provide the keys to future success.

Anticipating Second and Third Order Effects

Because strategy and armed conflict are so complex, any action has a multitude of second order effects (and third, fourth, and so on). Strategic decisions made today, particularly by the United States, will have second and third order effects on 21st century armed conflict. Some of these second order effects will be strategic and political. To take one example, by vigorously pursuing a revolution in military affairs designed to augment power projection and, perhaps, to lessen the need for allies, the United States may very well encourage the strengthening of regional security structures designed to minimize the need for American involvement or intervention. Many future innovations will bring equally unexpected second and third order effects. The development of military robotics, biotechnology, and psychotechnology, in particular, may unleash a hurricane of political, legal, and ethical problems.

CONCLUSION

No nation has ever undertaken a full revolution in military affairs unless it is responding to perceived risk or recent disaster. The paralysis of victory is great and vested interests always powerful. If historical patterns hold, the U.S. military may not be able to make the leap into the future on its own. It often seems that the Pentagon's plans for the future, including systems acquisitions, are based on "bygone battles." Ultimately, firm prodding may be necessary. This could come from Congress, the President and Secretary of Defense, or from battlefield defeat. If the nation is lucky, visionary leadership rather than American blood will inspire the necessary changes.

RECOMMENDATIONS

The key strategic challenges for the Army in the short- to mid-term (5 to 20 years) will be attaining greater strategic mobility, completing digitization, and becoming as effective at shaping the strategic environment as it is at responding to threats. The key strategic challenges for the mid- to long-term (15-30 years) will be:

- developing and integrating robotics and miniaturized systems;
- stressing the full modularity of equipment, systems and organization;
- developing methods for the rapid transformation of doctrine, concepts, and organizations; and,
- developing greater psychological precision, including the full integration of nonlethal capabilities.

To prepare for this second wave of transformation, the Army should use its futures-oriented programs and intellectual resources, particularly the Army after Next Project and the War College, to explore the strategic implications.

INTRODUCTION

The German philosopher Hegel held that revolutions are the locomotive of history. According to his theory, every social, political, and economic system builds up tensions and contradictions over time. Eventually these explode in revolution. Taking the argument one step further, Lenin held that a revolutionary did not have to wait for the explosion, but could speed it up, manipulate it, and control it. But Lenin was wrong. One cannot create a revolution in the way that an architect designs a building. Nor is it possible to control revolutions like a conductor leads an orchestra. Revolutions are much too big and complex for that. Those who live in revolutionary times can only make a thousand small decisions and hope that they move history forward in the desired direction. This holds as much for military revolutions as for any other kind.

A "revolution-centric" perspective on the development of war emerged among American strategic thinkers in the 1990s. Now security analysts, military leaders, and defense policymakers, not only in the United States, but around the world, accept the idea that some sort of revolution in military affairs is underway [1]. Its nature and eventual outcome, though, are less clear. One thing is certain: the United States has a greater stake in the revolution in military affairs than any other nation. By definition, revolutions upset existing relationships and hierarchies. Since the current configuration of global political, economic, and military power is favorable to the United States, the chances are that fundamental strategic change will prove deleterious to the American position. Washington is thus faced with the difficult task of modulating, directing, or controlling the revolution in military affairs.

History has seen two types of military revolutions. Operational and tactical revolutions occurred when new technology, operational concepts, or military organizations combined to generate a substantial increase in the effectiveness of military organizations. The revolution of the 1920s and 1930s that led to mechanized land warfare, strategic air war, and carrier war at sea is one example. Strategic revolutions have been much rarer. Alvin and Heidi Toffler suggest that strategic revolutions occur when a much broader shift in the method of production changes the entire panoply of human relationships, thus altering not only *how* militaries fight, but *who* fights and *why* they fight [2]. Today American strategic thinkers assume that the world is in the midst of an operational or technological military revolution and plan accordingly. In fact, a strategic revolution may be under way, spawned by and reflecting the information revolution.

Underestimating the extent of the ongoing revolution in military affairs and failing to understand its intricacies and second order effects can endanger American security. The need to think broadly and holistically is pressing. In simple terms, the information revolution is increasing interconnectedness and escalating the pace of change in nearly every dimension of life. This, in turn, shapes the evolution of armed conflict. Whether in economics, politics, or warfighting, those who are able to grasp the magnitude of this will be the best prepared to deal with it.

The architects of the 21st century American military must understand the broad political, economic, social, and ethical changes brought by the information revolution and by its manifestations—interconnectedness and an escalated pace of change. They must understand the effect these changes are having or might have on the evolution of armed conflict. Then, most importantly, they must develop some notion of what characteristics the future American military must have to prosper in the new strategic environment. The better an individual, an organization, or a state understands the nature of a revolution, the better its chances of emerging a winner. By examining the ongoing changes in the nature of armed conflict and thinking expansively, looking for wider implications and relationships, and exploring cross-cutting connections between technology, ethics, social trends, politics, and strategy, the architects of the future U.S. military can increase the chances of ultimate success. This study provides some suggestions on how this might be done.

PART I: STRATEGIC CONTEXT

INTERCONNECTEDNESS AND GLOBALIZATION

What is driving the current revolution in military affairs? Throughout history, many factors have altered the human condition: new ideas, religions, ecological shifts, disease, migrations, conquest, and so forth. Today technology, particularly information technology, is the locomotive, defining what is possible and pushing old ideas, values, methods, and organizations into obsolescence. As part of this, the information revolution is shaping the strategic environment in which armed conflict takes place. The revolution in military affairs is the dependent variable, driven and buffeted by wider changes. To understand future armed conflict, then, one must at least attempt to understand the political, economic, social, and ethical dimensions of the information revolution.

One of the most important changes associated with the information revolution is a dramatic increase in the interconnectedness of people around the world. This is evident at many levels and in many ways. For individuals, the number of people with whom they can cultivate some sort of relationship has increased exponentially. For most of human history, people only connected with the relatively few people who lived in their locale or whom they met on travels. Printing and literacy increased this somewhat by allowing people to develop at least a rudimentary understanding of others far away. Radio, the telegraph, the telephone, and television increased interconnectedness further by escalating the speed with which ideas could be transmitted and augmenting their psychological impact by making them more "human." Today, information technology allows the transmission of massive amounts of data to large audiences over great distances very quickly.

These relationships are much more dynamic, interactive, and powerful than the static one between author and reader.

On a personal level, individuals can cultivate a relationship with hundreds or thousands of people, whether through email, online chat, or other means. The explosion of wireless communications means that anyone who wants to can stay "connected" twenty-four hours a day. One can stand in the middle of an African game reserve many kilometers from the nearest paved road and read office email from a hand phone. By 2025, according to the United States Commission on National Security/21st Century, "the entire world will be linked, so that from any stationary or mobile station it will be physically possible to send and receive near-instantaneous voice, video and other serial electronic signals to any other station" [3]. As Bill Gates phrases it:

> Universal connectivity will bring together all the information and services you need and make them available to you regardless of where you are, what you are doing, or the kind of device you are using. Call it "virtual" convergence— everything you want is in one place, but that place is wherever you want it to be, not just at home or in the office [4].

Even more importantly, information technology allows everyone with access to it to become attuned to issues and problems in far-flung parts of the globe. There are tens of thousands of newspapers, newsletters, magazines, radio stations, and government documents available online. Chat rooms, email distribution lists, and online newsgroups exist for every conceivable topic. One can cultivate a fairly sophisticated understanding of any part of the world without leaving home. Information previously available only to those with the ability and the time to visit a library can now be delivered to anyone with a simple PC and a phone.

The information revolution opens new vistas for those who do leave home. For most of history, to migrate demanded extraordinary boldness or desperation.

Today, information technology allows potential migrants to both reconnoiter the area they would like to move to, and to retain reasonable ties with the family and friends left behind. International travel and migration—whether permanent or temporary—is thus easier and more common than it has ever been. The world is crisscrossed by networks, some based on ties like ethnicity or nationality, others on shared ideas, concerns, or ideology. These provide not only a source of information, but also a means to mobilize economic and political support for an organization or an idea.

Some dimensions of the information revolution and technological advancements are destabilizing or even dangerous. They have, for instance, blurred the distinction between fantasy and reality. Users of Internet Relay Chat (IRC) jokingly compare "RL" (real life) to the virtual world that entertains, informs, and, sometimes, confuses them [5]. The psychologically mature users understand the distinction between RL and IRC. For others, particularly adolescents and less mentally stable adults, the boundary is unclear, causing misunderstanding, confusion and anxiety. The fact that one can create an online persona unencumbered by reality can be liberating, but also dangerous for the immature or irresponsible. Freedom is always potentially dangerous. In this IRC is simply a microcosm of the wider problems brought by the information revolution. Advances in communication technology, especially the ability to meld reality and fantasy through things like morphing, when combined with the marketability of violent entertainment, confuse the young and the unstable who then feed each other's delusions via virtual communications. In the worst cases, fantasy and reality become hopelessly entangled and the result can be events like the murders at Columbine High School.

The information revolution has brought information overload. Everyone with a PC and an Internet connection runs the risk of being bombarded with ideas and images. While this can broaden an individual's perspective by providing access to different points of view and sources of information, it can also reinforce delusions by showing that others believe the same thing. Bizarre ideas and outright lies can be propagated much more easily than in the past. One has only to look at the plethora of conspiracy or racist web sites to see this at work [6]. And, information technology is also broadening the gap between the "haves" and "have nots," both within advanced societies like the United States and in the world as a whole.

Almost no dimension of modern life has been untouched by the information revolution. One of its most important effects has been the cascading globalization of economies. The "tactical" outcome is that businesses must have a global approach to markets, financing, trends, risk amelioration, partners, and suppliers. The "strategic" outcome is a linkage of economies around the world. "Economic downturns," notes the U.S. Commission on National Security, "that have usually been episodic and local may become, thanks to the integration of global financial markets, more systemic in their origins and hence more global in their effects" [7]. In a sense, this is not an entirely new phenomenon. Thomas Friedman points out that the period from the late 19th century to the middle of the 20th also saw substantial globalization driven by a decline in transportation costs arising from the invention of the railroad, steamship, and automobile [8]. But the process of globalization underway today is immensely more powerful in terms of its impact on politics, economics, culture, and values.

Every state must choose between participation in the globalized economy or persistent poverty. Participation means that the state—not just businesses within a state, but the government itself—must follow certain rules of behavior, including things like limiting corruption and making budgeting and finances transparent. "Transparency," write Robert Keohane and Joseph Nye, "is becoming a key asset for countries seeking investments. The ability to hoard information, which once seemed so valuable to authoritarian states, undermines the credibility and transparency necessary to attract investment on globally competitive terms" [9]. This has immense implications. Decisions made by multinational financial institutions, overseas banks, or investors on the other side of the world now determine the economic health of a nation nearly as much as decisions made by its own leaders. As Jessica T. Mathews writes, "National governments are not simply losing autonomy in a globalizing economy. They are sharing powers—including political, social, and security roles at the core of sovereignty—with businesses, with international organizations, and with a multitude of citizen groups, known as nongovernmental organizations" [10]. In a sense, all states have taken on some of the weakness, vulnerability, and lack of control that traditionally characterizes small states. As the ability of the state to control its economy fades, it is likely to become weaker across the board, thus leading to a major, perhaps revolutionary transformation of the global security system [11].

Economic globalization has a direct effect on security. In all likelihood, there will be states which refuse to participate. As they fall further and further behind, they may lash out with military aggression or terrorism. While Washington did not create globalization, Americans have been the most successful at adapting to it and thus have gained substantial advantages. "Those people who do not benefit from a more integrated global economy," according to the U.S. Commission on National Security/21st Century, "are unlikely to blame their own lack of social capital; they are more likely to sense conspiracy and feel resentment" [12]. In the eyes of many other nations, then, globalization is a deliberate strategy on the part of the United States to spread its influence and culture. While this is not true, the idea is pervasive and is likely to spark anti-American sentiment in states which come out losers during globalization. Mahathir Mohamad, Malaysia's Prime Minister, who accused the "Great Powers" of deliberately using globalization to cause his nation's 1997 economic crisis, is simply one of the first of what will be many leaders looking for a scapegoat to explain their shortcomings or frustrations [13]. As globalization erodes the ability of political leaders to fully control their own country's destiny, it simultaneously erodes their propensity to accept responsibility for events. This leads to a search for scapegoats. Often the symbols of globalization—the United States, the International Monetary Fund, and similar icons—will serve this function.

The information revolution, by eroding the control that authoritarian regimes can exercise over their citizens, is both liberating and destabilizing. The information revolution helped destroy Marxism-Leninism by stoking discontent and allowing opposition movements to form coalitions both within their states and outside. It may not necessarily represent the global ascendance of truth, but it certainly shortens the lifespan of lies. With the exception of dinosaurs like North Korea, Libya, Afghanistan, and Iraq which will tolerate opprobrium rather than surrender control of their citizens, world public opinion matters. Increasingly, states which practice repression do so through quick, spasmodic campaigns as in Rwanda. Since international intervention will continue to require a slow process of consensus building, the world will see a long series of humanitarian disasters in the face of rapid genocide or ethnic cleansing. In so many ways, the information revolution brings both good news and bad news, speeding the accumulation of information and, by increasing the data that must be considered and the range of available options, slowing the pace of decision-making.

While amplifying and magnifying connections, the information revolution has drastically increased the pace of change in human life. "By almost any measure," writes Hans Moravec, "the developed world is growing more capable and complex faster than ever before" [14]. Social, personal, economic, political, ethical, and technological factors all shift with breathtaking speed. Transformation and revolution are daily events. Successful individuals and organizations adapt to the pace of change and, at times, even control it. Those that cannot will face anxiety, stress, conflict, and failure.

Rapid change always has winners and losers, revolutionary change even more so. Much of the violence that will exist in the early 21st century will originate from the losers of the change underway today. The losers will be a polyglot group. They will include some societies or states unable or unwilling to adapt to globalization, particularly ones that cannot continence the lack of control and transparency that successful integration into the globalized economy demands [15]. The more benign ones may attempt isolation from the world (even given the human costs this will entail.) Others, like Iran or Afghanistan, will wrap their cause in cultural identity and use the tools of state power to resist or punish the United States, the International Monetary Fund, and other nations or organizations associated with globalization and interconnectedness. But there will also be losers within globalizing states. The protests against the World Trade Organization's 1999 Seattle meeting may give birth (or, at least, coherence) to a new ideology defined by opposition to globalization and interconnectedness. It is likely to bring together environmental activists, industrial workers, religious and cultural leaders opposed to the globalization, and political conservatives concerned about the erosion of national sovereignty and the intrusiveness of globalization [16]. This movement, with its dizzying, almost bizarre complexity and reliance on modern technology for mobilization and communication at the same time that it rejects the economic and social consequences of modernization, will typify many of the political movements of the coming era. Most of its components will not use violence and armed force, but some will. The information revolution will empower those opposed to it as well as those who accept it.

ORGANIZATIONAL CHANGE

The onward rush of information revolution is altering the shape of economic and political organizations. During the industrial age big, hierarchical organizations held advantage over smaller, less formally organized ones. Firms like Standard Petroleum and General Motors could crush or absorb smaller competitors through brute power. Small states, unless protected by some quirk of politics or geography, could seldom compete militarily with large ones. Today, the trend in the business world is toward macro-level integration and "strategic partnerships" but internal decentralization and the loosening of hicrarchies. Technology is forcing a major shift in paradigms of scale with adaptability and speed as important as aggregate resources [17]. By allowing multiple, cross-cutting connections between individuals and organizations, technology is dispersing power, creativity, and productive capability. Today, the successful commercial firm is one with a global perspective, a web of strategic partnerships, and internal flexibility based on project teams or work groups rather than hierarchies or bureaucracies. This phenomenon is migrating to the political world as well.

In the business world, the pressure to adopt modern organizational structure is a matter of institutional life or death. Corporations that resist risk failure. Governments, with their political and military resources, can hang on to outmoded structures longer than businesses. A government using outmoded organizational methods is in less danger of failure than a corporation that refuses to adapt. But clinging to old practices and organizations entails escalating costs and risks for governments as much as for corporations. As the same time that interconnectedness undercuts the viability of authoritarianism by allowing repressed citizens to communicate, organize, and mobilize, it also places handcuffs on elected governments. More and more, governments are blamed for economic and social conditions that they cannot ameliorate or control.

This reflects an historic deconcentration of political, economic, and ethical power. Carl Builder and Brian Nichiporuk wrote, "Since so many of the institutions of the nation-state are hierarchical and so many of the transnational organizations are networked, the net flow of power today tends to be out of the nation-state and into nonstate actors" [18]. Global public policy networks, which are loose alliances of government agencies, international organizations, corporations, nongovernmental organizations, professional societies, and other social groups, are becoming major political actors [19]. Information technology allows issue, goal, or project oriented networks to grow as dispersed actors communicate and coordinate across great distances, thus mobilizing pressure on governments [20]. Interest networks, if they have skilled leadership and an attractive "product," can wield influence disproportionate to their size. More and more, flexibility, creativity, astute marketing and responsiveness to supporters or constituents trump pure size or an aggregation of resources.

States are like dinosaurs toward the end of the Cretaceous Period: powerful but cumbersome, not yet superseded but no longer the unchallenged masters of their environment.

The information revolution is both a force for stability and for instability. On the positive side, it complicates the task of old-style repression and facilitates the development of grass roots civil society. It is not coincidence that there is more democracy today than at any time in history. But the information revolution also allows organizations intent on instability or violence to form alliances, thus making the world more dangerous. Some of the most complex struggles of the 21st century will pit polyglot networks against states. Colombia today offers a glimpse of this. There the alliance of political insurgents, drug cartels, international mafias, hired legal and economic advisers, and other affiliates is flush with resources and unbound by ethical or legal considerations. Characterized by "nimble new organizations" and "high tech gear," the Colombian drug traffickers contract out many functions, thus limiting the exposure of their core organization, and use the latest technology for encryption and cellular phone cloning [21]. The Revolutionary Armed Forces of Colombia, a leftist guerrilla movement which protects the heroin and cocaine industries, has amassed a small air force [22]. Such dangerous and polyglot enemies will probably propagate, posing great dangers for state security services. Hierarchies and bureaucracies face serious disadvantages when pitted against unscrupulous, flexible, adaptable enemies. If states are like dinosaurs, networks are like early mammals, still weak but waiting for the time that they will inherit the earth.

The strategic context in which future armed conflict will unfold will be a tempestuous blend of the old and the new. The information revolution is challenging the traditional frameworks which provided personal identity and moderated behavior, whether the family, village, church, place of employment, region, state, or nation. The replacements for these things are nascent, but not yet in place. As a result, the old bedrocks still matter—as Thomas Friedman points out, even the most forward-looking human still needs an "olive tree," which is his metaphor for "everything that roots us, anchors us, identifies us and locates us in this world" [23]. Individuals, organizations and states are redefining themselves, altering who they are, what they do, and how they relate to others. The world will never be the same.

THE CHANGING NATURE OF ARMED CONFLICT

The essence of warfare will always remain the same as antagonists attempt to impose their wills on each other while struggling with fog and friction. The information age, though, is generating important changes in the conduct of armed conflict. As these mature in the second decade of the 21st century and beyond, some will be "case specific," affecting a limited number of states or particular regions.

Others will be cross-cutting trends affecting nearly every participant in armed conflict and every mode of it. All organizations which participate in armed conflict, from the smallest terrorist cell to the most complex state military, are being changed by new technology, particularly information technology. For relatively simple war-making organizations, technology is helping to overcome shortcomings in communications, intelligence, and planning. For the complex militaries of advanced states, the change is even deeper, leading—at least according to American military thinkers—toward a fully "digitized" force where information technology eradicates fog and friction.

Other forces are also shaping armed conflict. The proliferation of weapons of mass destruction is particularly important. In fact, Martin van Creveld contends that proliferation will obviate traditional state-on-state war [24]. Even if one does not go that far, there is no question that proliferation will dramatically alter the strategic calculus for most nations. Nearly every moderately advanced state will have weapons of mass destruction, ballistic or cruise missiles, or the capacity to make them by the second or third decade of the 21st century [25]. This may not make armed conflict itself obsolete but, as van Creveld argues, will certainly make old-style major war unbearably dangerous.

Proliferation, in combination with interconnectedness and globalization, has created challenges to the political utility of armed force. This is likely to escalate in coming decades. This is not entirely new. History is replete with attempts to constrain, regulate, ban, or delegitimize armed force. The United Nations Charter, which constitutes binding international law for its signatory states, places strict limits on the conditions under which armed force is acceptable. Article 38 states:

> The parties to any dispute, the continuance of which is likely to endanger the maintenance of international peace and security, shall, first of all, seek a solution by negotiation, enquiry, mediation, conciliation, arbitration, judicial settlement, resort to regional agencies or arrangements, or other peaceful means of their own choice [26].

Since the end of the Cold War, challenges to the acceptability of armed force have continued and even accelerated, particularly in open political systems like the United States. To a large extent, this is a result of the information revolution. The casualties of war and their families are no longer faceless, but real, grieving humans. It is harder for policymakers to use force when their constituents understand the likely results. American leaders have responded by searching for modes of warfare that minimize friendly military and civilian casualties, particularly the use of precision aerial bombing. The problem is that such modes of warfare are inherently less decisive. It is possible that the concept of decisive victory will fade from the lexicon of strategy in coming decades. As Edward Luttwak contends, early 21st century war may look like early 18th century war where campaigns were waged for relatively limited objectives and the antagonists were not willing to pay a high blood cost for success [27].

Other elements of interconnectedness appear to be constraining at least state on state aggression. Globalization of the economy has created such tight linkages that armed violence in one part of the world has a ripple effect, often causing price increases or inflation elsewhere. This increases the pressure on hostile parties—particularly those integrated into the global economy—to refrain from war or seek a speedy end to one already underway. Undoubtedly there will be times when states consider the interests at stake in a conflict so important that they are willing to accept the costs of going to war. But the frequency of conflicts where a state sees its vital interests at stake and where war is seen as an acceptable means of promoting or protecting these interests is declining. This is particularly true for the United States. Preserving democracy and freedom against communism was a cause for which most Americans were willing to shed blood. Many of today's persistent conflicts, with their roots in ethnic and religious enmity, are difficult to understand and do not seem worth dying for, so minimizing casualties has become a central consideration for military planners, sometimes the preeminent one [28]. When the interests at stake are less than vital, the economic and political costs of armed conflict may serve as a brake. Ironically, though, these same constraints may prevent states from mobilizing and deploying overwhelming force in all but the most extreme cases, and thus cause those armed conflicts that do occur to be protracted. Again like wars in the early 18th century, early 21st century wars may drag on for extended periods of time.

In some ways, interconnectedness and globalization are creating new vulnerabilities for the United States. Future enemies are likely to have a better understanding of the American mentality than past ones and thus be able to craft more effective political and psychological campaigns. Their leaders may be attending college in the United States today. Those who are not can use the Internet as a window into the American psyche. And, as Martin Libicki suggests, small states may be able to use the "globalization of perception" to cast themselves as victims and mobilize world public opinion if they engage in conflict with the United States [29].

Interconnectedness also means that future enemies will have a potential constituency within the United States. This is not to imply that émigré communities are automatic breeding grounds for "third columnists." But immigrants or even native-born children or grandchildren of immigrants can, in some cases, retain a tie to their ethnic homeland which can lead them to lobby for or against American military involvement, as did Serbian Americans during the first stages of the 1999 air campaign. This increases the pressure on American policymakers and military leaders to minimize casualties if the use of force becomes necessary. Émigré communities can also provide logistics and intelligence support for terrorists. Interconnectedness will make protection against terrorism more difficult.

The U.S. Department of Defense and the military services hold that speed, knowledge, and precision will minimize casualties and lead to the rapid resolution of wars, thus minimizing the problems associated with the challenges to the political utility of force. States with fewer intellectual and financial resources than the United States will not have the luxury of using technology as a palliative for the strategic problems associated with interconnectedness and thus must seek other solutions. One such response has been renewed interest in multinational peacekeeping. The idea is that containing or deterring armed conflict limits the chances of full blown war. Some states may turn instead to strategies of passive defense. One of the dilemmas of interconnectedness is that what happens in one place affects many others, but explaining this to mass publics remains difficult. Aggressive states or non-state actors will also have to find ways to transcend the constraints brought on by interconnectedness. Some will rely on proxy conflict, providing surreptitious or, at least, quiet support to insurgents, militias, or terrorists whose activities further the aims of the sponsoring state. Some may attempt hidden or camouflaged aggression, particularly cyberwarfare aimed at the information systems of their enemies. Some—particularly those which find their ambitions blunted by the United States—will turn to political methods, ceding battlefield superiority to the American military while seeking to constrict Washington through legal and political means. America's military advantages, after all, are not always matched by an equal political and diplomatic superiority.

Because globalization and interconnectedness erode the control which regimes can exercise within their states, those with a shaky hold on power will often seek scapegoats but will sometimes turn to the time-tested method of solidifying internal unity by external aggression as well. Since globalization and interconnectedness raise the political and economic cost of protracted war, regimes which seek to deflect internal discontent through external aggression will probably seek lightening campaigns, seizing something before the international community can reach consensus on intervention. Future actions like the Iraqi seizure of Kuwait are not out of the question, at least for states which believe that the United States cannot or will not stop them. Whether the United States can be deterred from intervention by weapons of mass destruction or terrorism is one of the central questions for the future global security environment.

PRIVATIZATION

Interconnectedness, the dispersion of power and knowledge that flows from the information revolution, and the eroding legitimacy of armed force are leading toward privatization in the realms of security and armed conflict. This has a long history, particularly functions involving technical skills beyond those of the average warrior. During the early modern period, for instance, artillery and siege engineering were often handled by contractors rather than regular soldiers. Today, as warfighting becomes ever more complex and the costs of training and retaining technical specialists escalate, the same process is occurring. Within the United States, many jobs done by uniformed personnel a few years ago are now handled by contractors. This includes not only administrative tasks but, increasingly, planning, analysis, war gaming, training, and education. To take one example, the United States recently established the African Center for Strategic Studies (ACSS) to help African states improve their civil military relations and their ability to understand national security planning and defense budgeting [30]. This is similar to Department of Defense schools established for Europe, the Asia-Pacific region, and the Americas. But unlike these others, which are operated by the U.S. military, the corporation Military Professional Resources International, which is composed mostly of retired U.S. military officers, is responsible for the development and implementation of the curriculum for ACSS (with oversight from the Office of the Secretary of Defense) [31]. Similarly, contractors play a vital role in most Department of Defense and service wargames. As in the business world, "outsourcing" allows the U.S. military to acquire expertise while retaining organizational flexibility.

Today the contracting out of military functions is most pervasive in the United States. In coming decades, other states will probably turn to it both as a means of acquiring cutting edge expertise and providing surge capacity during major operations. They might, for instance, hire medical support when they go to war rather than building an extraordinarily expensive military infrastructure. Privatization will give many state militaries and non-state actors the ability to acquire advanced skills much more effectively and quickly than if they had to develop them internally. Drug cartels and rogue states, for instance, might simply hire the best available information warfare experts. This could decrease the qualitative advantage held by the United States and other advanced militaries, at least in key areas where the expense of contracting is warranted. The same could happen in the realm of combat itself. The world is witnessing the re-emergence of powerful and effective mercenary firms, particularly in places like Sub-Saharan Africa where state militaries are rife with problems and weaknesses. The best known was a company called Executive Outcomes which was composed of combat veterans from the ex-South African Defence Force. This company not only offered military advice and consulting, but also combat forces which saw action in Angola and Sierra Leone [32]. While Executive Outcomes officially closed shop at the end of 1998 (largely in response to South Africa's passage of the Military and Foreign Assistance Act), a successor or successors may emerge [33]. In fact, there were reports in early 1999 that South African mercenaries simply relocated to Eastern Europe and continued to supply the Angolan rebels (who could pay with the proceeds of diamond sales) [34]. This is simply the starkest example of a wider trend toward the privatization of security [35].

As nations seek ways to attain a surge capacity without the expense of sustaining a large, peacetime military, and as they face difficulties recruiting from their own populations, contracting will be an attractive option for filling the ranks. Eventually, advanced nations like the United States may replicate the development of the Roman army. During the early days of the Roman Republic, the army was composed largely of citizen soldiers who served during times of threat. Eventually this gave way to an army of long-service professionals attracted by the financial benefits service could provide. By the late imperial period, it was increasingly difficult to recruit Romans because of other economic opportunities and because the prestige of military service declined. At that point, the army was composed mostly of foreigners attracted by the chance to gain citizenship and other material inducements. There is the possibility that the future U.S. military may have to turn to foreign recruits in order to fill its ranks. This is simply one additional form of the privatization of security.

History suggests another twist that privatization might take as well. Whenever rich, powerful companies believed that no state was willing to shed blood to defend their people and assets, the temptation was to form private armies and navies. The British East India Company, for instance, once had one of the largest military establishments on earth. If coming decades see the development of truly transnational or non-national corporations, this process may be repeated. Corporate armies, navies, air forces, and intelligence services may be major actors in 21st century armed conflict. This will open new realms of strategy and policy. Would it, for instance, it be legal and acceptable for the United States to declare war on a corporation that was guilty of armed aggression against a friend? To sign an alliance with one?

ASYMMETRY

States which decide to commit aggression in coming decades will know that if the United States and the world community decide to counter the aggression, they can. The qualitative gap between the U.S. military and all others is wide and growing. This leaves aggressors two options: they can pursue indirect or camouflaged aggression, or they can attempt to deter or counter American intervention asymmetrically. While the word "asymmetry" only recently entered the American strategic lexicon, the idea is not new. From Sun Tzu's contention that "all warfare is based on deception" through B.H. Liddell Hart's advocacy of the "indirect approach" to Edward Luttwak's "paradoxical logic of strategy," strategic thinkers have long trumpeted the wisdom of avoiding the enemy's strength and probing for his weakness [36]. Asymmetry simply means making maximum use of one's advantages. It is the core logic of all competitive endeavors, whether sports, business, or war. Consistent winners master this logic.

Through what might be called "low" asymmetry, militaries facing a superior opponent avoid open, force-on-force battles and rely on hit-and-run tactics, deception, camouflage, dispersion, the use of complex terrain like cities, mountains, and jungles, guerilla warfare, or terrorism. They often drag out the conflict, playing on an asymmetry of will or patience, and make use of their own tolerance for pain and cost. Throughout history, low asymmetry has allowed the weak to overcome the mighty, from the defeat of Darius by Scythian guerrillas through the American Revolution and Spain's expulsion of Napoleon to the 20th century wars of liberation in Algeria, Zimbabwe, Namibia, Vietnam, and other colonies. By contrast, "high" asymmetry is favored by militaries facing an enemy which outnumbers them or in situations where casualties must be minimized. High asymmetry uses superior technology, information, training, leadership, and the ability to plan and coordinate complex operations to overcome quantitative disadvantages or limit the blood cost of warfare. Many colonial wars, from those of Caesar through the campaigns of the Spanish conquistadors to the European conquest of Africa in the 20th century evinced this type of asymmetry. Battles like Marathon, Agincourt, Blood River, and Omdurman were won by asymmetry. In the modern context, *blitzkrieg*, whether used by its German architects or by the coalition forces expelling Iraq from Kuwait, is an example of high asymmetry.

Asymmetry is a characteristic of periods of rapid change, particularly revolutionary ones. In geological history, there have been times when many new species emerged. Most proved unable to survive, leading to new periods with less diversity. Military history follows the same pattern: periods of great diversity follow periods of relative homogeneity. The current era is one of diversity. In coming decades, some methods of warfare or of military organizations will prove dysfunctional, thus leading to greater homogeneity. But for the period of diversity, asymmetry will be a dominant characteristic of armed conflict.

COMBATANTS

Throughout the 20th century both states and nonstate actors have undertaken armed conflict. While some nonstate actors, particularly insurgent movements, have shaped history, state combatants have been the most significant. Great wars tend to work against diversity in methods and organizations for armed conflict, serving to weed out the dysfunctional from the successful. The great wars of the 20th century did precisely that for state combatants. While there certainly was great variation among states in terms of the size, effectiveness, and technological advancement of their militaries, there were significant similarities in terms of military organizations and methods. These included: (1) hierarchical organization into services defined by the primary operating environment, and into discrete groups of officers and enlisted personnel; (2) formal, hierarchical procedures for planning and decision-making; (3) a professional core of some type reinforced, in many cases, by a reserve; (4) emphasis on linear operations (supported, in some cases, by nonlinear special operations) organized into battles, campaigns, and wars; and, (5) reliance on the equipment produced by advanced industry and science, and on formal supply systems. Nonstate combatants varied from this. Their organization tended to be less formal, with some combination of guerrilla combatants, political cadres, terrorist cells, and militias. Their operational techniques stressed hit and run tactics, harassment, psychological actions, and guerrilla activities, often using complex terrain. Their supply systems tended to be a blend of the formal and informal, often relying on captured arms, ammunition, and equipment, in large part because they did not have the geographic, financial, or organizational resources to do otherwise.

Often the ultimate objective of nonstate combatants was to take on the characteristics of state ones.

In the opening half of the 21st century, the types of state and nonstate combatants which have characterized recent armed conflict will continue to exist, but they are likely to be joined by new forms. The U.S. military probably will be the first *postmodern state* combatant, attaining greatly amplified speed and precision by the integration of information technology and development of a system of systems which links together methods for target acquisition, strikes, maneuver, planning, communication, and supply. Its organization will be less rigidly hierarchical than that of modern state combatants. This will both reflect the fact that a digitized force needs less rigidly centralized control, and that the sort of high tempo, pulsed, holistic, nonlinear operations it will undertake simply will not work with rigid, centralized control [37]. The final type of combatants in 21st century armed conflict are likely to be *postmodern nonstate* ones. This will consist of loose networks of a range of nonstate organizations, some political or ideological in orientation, others seeking profit. They will work toward an overarching common purpose, but will not be centrally controlled or have a single center of gravity.

When one type of combatant fights a similar type, the result will be a more or less symmetric. Even though one side may prove more capable or competent than the other, their basic tactics, strategies, and weapon systems will be similar. But much of 21st century armed conflict will be distinctly asymmetric, pitting one of the four types against a different one. In all asymmetric conflicts, the combatant at a material disadvantage will succeed only when it can make use of greater will and creativity. When there is no asymmetry of will and creativity, postmodern state combatants will generally have an advantage. When there is an asymmetry of will and creativity, anything is possible.

PART II: IMAGES OF FUTURE WAR

THE SERVICE AND DOD VIEW

The specific shape of future armed conflict will be determined by policy decisions, technological developments, economic, political, and social trends, and by the geostrategic configuration that emerges. This dizzying complexity makes it impossible to predict the path of future warfare with certainty. At best, images can be sketched. Broadly speaking, the opening decades of the 21st century are likely to see some combination of three modes of warfare: formal war, informal war, and gray area war.

Formal war pits state militaries against other state militaries. Since the 17th century, it has been the most strategically significant form of armed conflict and will probably remain so for at least a few more decades, perhaps longer. For this reason, it has been the focus of most futures-oriented thinking within the U.S. military and Department of Defense. American policymakers and military leaders are attempting to define and create the first postmodern state military, primarily for use against "rogue states" or a "near peer competitor" that might appear early in the 21st century [38].

The official vision of future war reflects the belief that "information superiority" will be the lifeblood of a postmodern military and thus the key to battlefield success. According to Secretary of Defense William Cohen,

"The ongoing transformation of our military capabilities—the so-called Revolution in Military Affairs (RMA)—centers on developing the improved information and command and control capabilities needed to significantly enhance joint operations" [39]. Deriving from a "system of systems" that connects space-based, ground-based, and air-based sensors and decision-assistance technology, information superiority—should it be realized—would allow American commanders to use precision weapons—many fired from safe locations far from the battlefield—to strike the enemy's decisive points at exactly the right time [40]. The idea is that American forces will be nearly omniscient while enemy forces are confused and blind [41].

The most important expression of the official American vision of future war is a document known as *Joint Vision 2010* [42]. Known within the Department of Defense as "JV 2010," this is the "conceptual template" for the future U.S. military able to attain "full spectrum dominance"—qualitative superiority over any anticipated enemy in any anticipated operating environment. JV 2010 holds that the key to success in an increasingly lethal battlespace will be "dominant battlespace awareness" growing from the system of systems. This will allow the postmodern U.S. military to survive on a battlefield replete with weapons of mass destruction and precision guided munitions. JV 2010 states:

> To cope with more lethal systems and improved targeting, our forces will require stealth and other means of passive protection, along with mobility superior to the enemy's ability to retarget or react or our forces. Increased stealth will reduce an enemy's ability to target our forces. Increased dispersion and mobility are possible offensively because each platform or individual warfighter carries higher lethality and has greater reach. Defensively, dispersion and higher tempo complicate enemy targeting and reduce the effectiveness of area attack and area denial weaponry such as weapons of mass destruction (WMD). The capability to control the tempo of operations and, if necessary, sustain a tempo faster than the enemy's will also help enable our forces to seize and maintain the initiative during military operations [43].

As the U.S. military evolves along the lines described in JV 2010, it will gradually abandon old operational concepts like massed force and sequential operations in favor of massed effects and simultaneous operations. These will be possible because information technology will allow commanders to identify targets and coordinate complex actions much better than in the past. In addition, technological advances, according to JV 2010, "will continue the trend toward improved precision. Global positioning systems, high-energy research, electromagnetic technology, and enhanced stand-off capabilities will provide increased accuracy and a wider range of delivery options" [44].

To make maximum use of emerging technology, JV 2010 outlines four new operational concepts to guide the development of U.S. armed forces and military strategy: *dominant maneuver* which is defined as "the multidimensional application of information, engagement, and mobility capabilities to position and employ widely dispersed joint air, land, sea, and space forces to accomplish the assigned operational tasks"; *precision engagement* which will allow accurate aerial delivery of weapons, discriminate weapon strikes, and precise, all-weather stand-off capability from extended range; *full-dimensional protection* of American forces based on active measures such as battlespace control operations to guarantee air, sea, space, and information superiority, and integrated, in-depth theater air and missile defense, and passive measures such as operational dispersion, stealth, and improved sensors to allow greater warning against attack, including chemical or biological attack; and *focused logistics* which is "the fusion of information, logistics, and transportation technologies to provide rapid crisis response, to track and shift assets even while en route, and to deliver tailored logistics packages and sustainment directly at the strategic, operational, and tactical levels of operations" [45].

Joint Vision 2010 was intended to synchronize the futures-oriented programs which the services had begun to develop. Where JV 2010's time frame was mid-term, the Joint Experimentation Program created in 1998 at the United States Atlantic Command (USACOM, since renamed U.S. Joint Forces Command or JFCOM) sought to expand the U.S. military's thinking about future warfare by weaving together the services' futures programs [46]. This is a very ambitious undertaking. Futures-oriented thinking deals with force development which is a responsibility of the services. In fact, most of the futures thinking within the U.S. military is still done by the services. The Army, the Air Force, and the sea services have each developed a range of futures programs based on their expectation about the future security environment and the future of war.

So far, the Army's program is the most elaborate. Since there is no White House, National Security Council, or congressional concept of the future security environment or long-term American national security strategy, the Army, like the other services, has had to craft its notion of the future role of landpower on its own [47]. It has formulated a vision that is highly innovative in its approach to technology, organization, and leadership, but conservative in its assumptions about the nature of warfare and the purposes of American military power. This blend of innovation and conservatism runs throughout the documents and programs that explain the Army's view of the future.

Army Vision 2010, which explains how the Army will support the ideas introduced in *Joint Vision 2010*, argues that landpower will remain the most salient form of military power in the future security environment because many American military operations will fall on the lower and middle portions of the continuum of military operations, because most foreign militaries will remain landpower oriented, and landpower makes permanent "the otherwise transitory advantages achieved by air and naval forces" [48]. *Army Vision 2010* also argues that the Army is best suited among the services to deal with asymmetric challenges such as urban combat, terrorism, information warfare, and insurgency. While it notes that operations other than full-scale war will be the most common task of the 21st century Army, it identifies the possibility of conventional war against "once dominant states [which] perceive an unfavorable shift in power relative to their neighbors." Oil and "radical fundamentalism," according to *Army Vision 2010*, might motivate war in the "Euro-Middle East region," while a shortage of food and arable land might do likewise in "the Asian arc." Should either of these happen, the U.S. Army might be called on the defend or liberate territory, contain the conflict, or perform other missions [49].

To transform the concepts outlined in documents like *Army Vision 2010* into reality, the Army developed a series of battlelab simulations and exercises called Louisiana Maneuvers [50]. Begun in 1992, this quickly grew into the elaborate "Force XXI" process that uses battle laboratories, warfighting experiments, and advanced technology demonstrations to generate and test ideas [51]. In the mid-1990s, Army Chief of Staff General Dennis Reimer decided that his service needed to look even deeper into the future. The pace of change in the modern world had become so intense, General Reimer concluded, that complex organizations like the Army must extend their strategic planning horizons. And the main weapon platforms of the Army, including the Abrams main battle tank, the Bradley fighting vehicle, and the Apache attack helicopter were expected to approach obsolescence around 2015. General Reimer thought it necessary to craft a rigorous method to decide whether the Army should seek a new generation of tanks, fighting vehicles, and helicopters or instead pursue "leap ahead" technology.

The framework for this analysis is the Army After Next Project—an ongoing series of wargames, workshops, studies, and conferences which explore the feasible strategic environments of the 2020-2025 period and speculate on the sort of technology, force structure, and operational concepts that the U.S. Army might need [52]. One of the most crucial parts of the Army After Next process has been identifying the most likely or dangerous type of enemy. *Speed and Knowledge*, which was the first annual report of the Army After Next Project, singled out a "major military competitor" [53]. This would be a nation-state that threatens the United States or U.S. interests but cannot or does not emulate the digitized American military. Such an enemy would attempt to offset technological inferiority with relatively cheap counters such as land and sea mines, distributed air defense, coastal seacraft, submarines, inexpensive cruise and ballistic missiles, and unsophisticated weapons of mass destruction which have become, as Richard Betts points out, weapons of the weak rather than the most advanced [54]. Quantity would substitute for quality. The Army

After Next Project seeks to design a force with superior operational and decisional speed, strategic mobility, and battlefield awareness to defeat such a "major military competitor."

The Army After Next Project assumes that precision weapons will make the battlefield of 2025 so deadly that the defensive will be strengthened, making extended maneuver possible only when the enemy's advanced systems have been degraded and when one's own forces have very high degrees of mobility and speed. Mobility and speed will allow distributed, decentralized, high tempo operations with what are described as "cascading" effects. "Tactical success," according to the second annual report of the Army After Next Project, "piled up nearly simultaneously across the entire battlespace, could then lead under the right circumstances to rapid operational-level disintegration as the enemy's plans are first foiled and then shattered—even as his ability to control his own forces evaporates before he can respond" [55].

The Army After Next will be built on knowledge accruing from advanced information technology, specifically an integrated, multilayered system of systems that fuses information from a variety of sources and provides "a coherent, near real time, common picture of the battlespace." The Annual Report states that "knowledge is paramount. . . . the unprecedented level of battlespace awareness that is expected to be available will significantly reduce both fog and friction." It continues:

> Knowledge will shape the battlespace and create conditions for success. It will permit...distributed, decentralized, noncontiguous operations...It will provide security and reduce risk. Through the identification of enemy strengths, weaknesses, and centers of gravity, coupled with near complete visibility of friendly force status and capabilities, knowledge will underwrite the most efficient application of all elements of military power—enabling higher tempos of operations. Knowledge will also focus and streamline the logistics support required to maintain high tempos [56].

Organizationally, the Army After Next Project anticipates a hybrid U.S. Army combining very advanced components with "legacy" forces.

This will include: contingency forces including Battle Forces, Strike Forces, Campaign Forces, Homeland Defense Forces, and Special Forces [57]. Through this combination, the future U.S. Army would retain flexibility and be able to operate in coalition with allies who had not built "digitized" forces. Throughout the Army After Next Project's studies, programs, wargames and seminars, though, emphasis remains on countering cross-border aggression against a state where the United States had economic interests (usually petroleum) by another state using combined arms warfare with a few additional technological twists and capabilities. Invariably, the "blue" forces emerge victorious leaving the Army unprepared to think about the consequences of or responses to defeat.

The U.S. Air Force's vision of future war is also characterized by a combination of creativity and conservatism. The Air Force 2025 study, commissioned by the Chief of Staff of the Air Force, was a cauldron of new, creative thinking. It solidified the position of the Air University as the U.S. military's cutting edge source of ideas. Often using teams with a senior researcher of colonel or lieutenant colonel rank and a number of majors, Air Force 2025 explored topics such as information warfare, unmanned aerial combat platforms, organizations to deal with the gray area between peace and war, and ways to most efficiently erode an enemy's unity and will [58].

To some extent, the Air Force is more open to innovative strategic concepts than the other services, particularly the Army and the Navy. The *Air Force Strategic Plan* notes that exotic technologies such as micro-technology, biotechnology, and nanotechnology could alter the shape of future battlefields. But generally, Air Force's senior leaders see future warfare as an extrapolation of the 1990s. The *Air Force Strategic Plan* indicates that non-state enemies and asymmetric strategies will pose challenges and the U.S. military must become more proficient in environments like the infosphere, space, and urban areas, but assumes continuity in American strategy and in the overall nature of armed conflict. Ironically, the Air Force planning document notes the ongoing diffusion of information technology and the commercialization of space, but does not suggest that these might challenge the notion of "information superiority" on which *Joint Vision 2010* is built.

The sea services also subscribe to the notion that future warfare will be a high-tech version of late 20th century combat. But the Marines, at least, are looking seriously at fairly radical changes in tactical and operational procedures, including new organizations and doctrine. In fact, the Marines are in many ways the service most amenable to true transformation. For instance, the Marine Corps After Next (MCAN) Branch of the Marine Corps Warfighting Laboratory is exploring what it calls a "biological systems inspiration" for future warfighting. According to its web site:

> . . . for the last three centuries, we have approached war as a Newtonian system. That is, mechanical and ordered [sic]. In fact, it is probably not. The more likely model is a complex system that is open-ended, parallel, and very sensitive to initial conditions and continued "inputs." Those inputs are the "fortunes of war." If we assume that war will remain a complex and minimally predictable event, the structures and tactics we employ will enjoy greatest success if they have the following operational characteristics:
>
> - dispersed
> - autonomous
> - adaptable
> - small
>
> The characteristics of an adaptable, complex system closely parallels biology. For that reason, much of the efforts of MCAN focus on exploiting biological inspiration in future military systems [59].

To move in this direction would require technology like biomimetic engineered materials; small, "bug like" robotics; neural or neuronal nets capable of complex, adaptive responses; parallel computers; and, nanotechnology. But there is more to it than that. What the Marine Corps After Next group is grappling with—to use a phrase that is close to becoming a cliché—is a "paradigm shift." The futures-oriented programs of the others services focus more on "paradigm refinement"—doing what they have traditionally done better through new technology and its associated concepts and organizations. The real issue becomes whether the Marines can truly undertake a paradigm shift while the other services, the Department of Defense, and some of the key leaders of the Marine Corps concentrate on paradigm refinement.

The end of the Cold War largely eradicated the primary mission of the Navy: retaining control of the seas in the face of enemy sea power. In response, the Navy has shifted its focus from fleet encounters and protection of sea lines of communication from hostile forces to influencing events on land via Marine Corps operations and strikes launched from the sea [60]. To do this, the Navy plans to continue using existing weapons platforms, particularly carrier battlegroups, surface platforms, and multi-purpose submarines [61]. It talks of decisive victory in future war using cruise missiles, naval aviation, and better target acquisition [62]. The Navy holds that "sea strike"—attacking targets on land from the sea, is a "revolutionary" concept [63].

Because of the massive cost of a ship, the Navy concentrates more on applying new technology to existing or proposed ones rather than the development of whole new weapons platforms as the Air Force and Army prefer. What this means is that great efforts are going to have to be made to protect things like surface ships that emit an immense electronic signature, particularly as more states develop precision weapons, weapons of mass destruction, and improved means of target acquisition through things like the purchase of commercial overhead imagery. While many theorists contend that "if it can be found, it can be destroyed" is one of the "rules" of the current revolution in military affairs, the Navy assumes that this is either not true, or will not hold for America's enemies. To a great extent, this is one more illustration of the hubris that pervades the official American perspective on future warfare. Unassailable American technological superiority and "full spectrum dominance" are articles of faith.

Like the Army and Air Force, the Navy is exploring a different approach to warfare (albeit using existing platforms). In the case of the

Navy, this is called "network-centric warfare" in which a postmodern military using networked sensors, decision makers and shooters collapses an enemy's will to resist quickly and efficiently [64]. According to Vice Admiral Arthur K. Cebrowski, President of the Naval War College, network-central warfare, "enables a shift from attrition-style warfare to a much faster and more effective warfighting style characterized by the new concepts of speed of command and the ability of a well-informed force to organize and coordinate complex warfare activities from the bottom up" [65]. A military which masters network-centric warfare, according to its adherents, will achieve information superiority, reach out long distances with precision weapons, and collapse an enemy's will through the shock of rapid and closely linked attacks.

Elsewhere within the Department of Defense, the search continues for ways of applying new technology to traditional modes of armed conflict. The joint experimentation program at JFCOM is an important part of this. It remains to be seen, though, whether this will incorporate analysis of true paradigm shifts in addition to paradigm refinement, and whether the results of the experiments will have a meaningful effect on the services, the Department of Defense, Congress, and other national political leaders at least as long as the threat to American security remains manageable.

Other pockets of innovation and creativity exist through the Department of Defense. For instance, the Pentagon's Office of Net Assessment, which was the birthplace of American thinking on the revolution in military affairs, has developed an Operational Concepts Wargaming Program to explore the ideas outlined in JV 2010 [66]. The Defense Science Board has done some useful thinking about a new land-based military unit which reflects the operational preferences and technological capabilities of a postmodern military. This new unit would be light, agile, and potent. It would operate in a distributed and desegregated fashion, utilizing high situational awareness generated by information technology, depending on remote fires, connected by a robust information infrastructure, and supported by precision logistics [67]. Such an organization could provide a rapid intervention capability and prepare the way for heavier units which would arrive later. It would fight for two weeks or less and then either be reinforced or withdrawn. The basic element would be "combat cells" which would make extensive use of unmanned vehicles and robotics, using humans "only when necessary." They would avoid direct firefights, remaining dispersed most of the time for survivability, massing only to repulse a major attack. Information technology would be central: "A key capability for combat cell mission success is maintaining a local awareness bubble larger than the enemy's" [68].

Along similar lines, a study group at the Department of Defense's Center for Advanced Concepts and Technology has explored the concept of "rapid dominance" attained by "shock and awe." This is a very important attempt to integrate a psychological dimension into mainstream thinking on the revolution in military affairs. The goal is to use a variety of approaches and techniques to control what an adversary perceives, understands, and knows [69]. To do this, a rapid dominance military force must have near total or absolute knowledge and understanding of itself, the adversary, and operational environment; rapidity and timeliness in application; operational brilliance in execution; and near total control and signature management of the operational environment.

It is not clear, though, what effect an inability to attain one or more of these things might have on a postmodern military. While attaining a perfect picture of the battlefield would give the U.S. military great advantages, reliance on this would also be a vulnerability. Might the future U.S. military become so accustomed to the absence of the fog of war that it could not overcome imperfect knowledge when it does occur? As one dimension of the paradoxical logic of strategy, weakness sometimes begets strength and strength sometimes begets weakness. Eventually, this intricate conundrum might erode the battlefield advantage of the American armed forces.

All of the services agree that the future U.S. military needs some sort of highly capable, rapidly deployable expeditionary unit. The core concept behind this is "strategic preclusion" which, in a crisis, would allow the U.S. military to achieve battlefield dominance before an enemy has completed "operational set" [70]. This would force the opponent to either concede or face inevitable defeat. Again, the expectation is that future warfare will be a reprise of *Desert Shield/Desert Storm*—unambiguous, cross-border aggression by one state against another. The services, however, offer few explanations of why American political leaders would use military force early in a crisis when they traditionally consider it a last resort. Similarly, there is little indication of how the various future strike and expeditionary forces might be used against nontraditional enemies or ambiguous aggression. "Strategic preclusion" may be an example of the tendency to prepare to fight the previous enemy rather than future ones.

The official vision of information warfare follows a similar logic. Joint doctrine defines information operations as "actions taken to affect adversary information and information systems while defending one's own information and information system" [71]. Despite immense debate within the services and the Department of Defense, the general notion is that information is an "enabler" of traditional forms of military activity, "an amalgam of warfighting capabilities integrated into a CINC's theater campaign plan. . . ." [72]. While official thinking accepts the fact that information technology has had a revolutionary effect, this revolution is thought to have cemented the strategic realities of the past, particularly the technological advantage held by the U.S. military rather than creating new vulnerabilities or the potential for enemies to match or surpass the United States. The American technological advantage is an article of faith in official thinking, largely because of the extent of investment and effort devoted to it. Little consideration is given to the creativity which might be born from the desperation of America's enemies.

The official American view of the future consistently treats technology, particularly information technology, as a force multiplier rather than as a locomotive for revolutionary transformation. Concepts such as "strategic preclusion," "full spectrum dominance," and "information superiority" reflect the situation of the 1990s—a qualitatively dominant U.S. military focusing on deterring or defeating traditional cross-border aggression. Most official documents accede that future enemies will attempt asymmetric methods, but it is what might be called a "moderate" asymmetry rather than a radical type. Official discussions of technologies that appear to have the potential to be truly transformative—nonlethal weapons, strategic information warfare, robotics, and so forth—are conservative, seeing these things as support systems in conventional warfighting rather than new modes of warfare [73]. With the exception of adding three new tasks for the U.S. military—space operations, information warfare, and homeland protection—the official vision anticipates few if any strategic shifts. 21st century war, according to official thinking, will be mirror late-20th century war, with new technology allowing future generals and privates to do what past warriors could only dream of.

ASYMMETRY AGAIN

The notion that 21st century warfare will pit an omniscient postmodern U.S. military in lop-sided, lightening operations against evil aggressors is enticing. But is it accurate? Perhaps, particularly in those instances where an aggressor does not expect American involvement. There may be times when the United States surprises an aggressor using Soviet-style equipment, tactics, and operations. Such wars would be a reprise of *Desert Storm*. Opponents who *do* anticipate and plan for American involvement, though, are likely to attempt to counter the prowess of the U.S. military through asymmetric means.

To some extent, current official thinking recognizes this. In fact, asymmetry has become a central concept in official American thinking about future warfare. While *Joint Vision 2010*, which was released in 1996, does not explicitly mention asymmetry or asymmetric counters, all key planning documents now do. The Air Forces' *Global Engagement* notes that "hostile countries and non-state actors [will] seek asymmetric means to challenge US military superiority"; the 1998 *Annual Report of the Army After Next Project* contends that "major competitors will probably develop creative *asymmetric strategies*"; and the 1999 Joint Strategic Review provides an in-depth analysis of the implications of asymmetric methods. The reason is fairly simple: the Gulf War seemed to show that the United

States cannot be defeated by conventional Soviet-style methods. If anything, the gap between the American military and opponents who might attempt force-on-force combat in open terrain is growing. No potential enemy will soon undergo an information-based revolution in military affairs and develop a postmodern force. But enemies still feel the need to challenge the United States or, at least, to make themselves impervious to American intervention.

The question then becomes: what forms of asymmetry will be most common and, more importantly, most problematic for the United States? Enemies using precision munitions or weapons of mass destruction to complicate deployment into a theater of operations could pose a serious challenge to some of the most basic tenets of American strategy [74]. Since the campaigns of Ulysses Grant and William Sherman, the "American way of war" has called for the build-up of massive amounts of materiel and supplies in a theater of operations, and then the use of this material advantage to attain decisive victory through a strategy of annihilation [75]. This is contingent on the enemy's absence of effective power projection to strike at the rear bases. In the American Civil War, the Confederacy simply did not have the force necessary to capture Union depots at places like City Point, Virginia. In the European theater of World War II, the English Channel, the Royal Air Force, and the Royal Navy kept the rear bases safe until adequate American forces were deployed. And, in the Gulf War, American airpower and landpower protected the rear bases.

In a future where enemies have some precision guided munitions and weapons of mass destruction (along with delivery systems), in-theater sanctuaries may not exist. Even air superiority and theater missile defense would be inadequate against a nuclear-armed enemy, since they cannot assure the sort of 100 percent effectiveness that is necessary. Given this, the future American military may confront an enemy using a counter-deployment strategy in which precision guided munitions and ballistic missiles, whether with nuclear, biological, and chemical warheads or conventional ones, are used to attack U.S. bases and staging areas both in the United States and in a theater of operations, and to threaten states that provide support, bases, staging areas, or overflight rights to the United States.

An enemy using a counter-deployment strategy would have to be met with a combination of strategic airpower, naval strike forces, theater air superiority, theater missile defense, focused logistics to minimize the supplies needed in theater, and a range of methods to limit the need for a lengthy build-up of forces, equipment, and supplies. As the 1997 National Defense Panel wrote, "The days of the six-month build-up and secure, large, rear-area bases are almost certainly gone forever. WMD will require us to increase dramatically the means to project lethal power from extended ranges" [76]. The capacity to deploy forces and resupply them directly from the continental United States into a theater of operations could prove invaluable, minimizing the chances that states in the theater of operations could be coerced into denying U.S. forward bases or staging areas.

The need to protect U.S. forces from strikes launched by an enemy using a counter-deployment strategy suggests the need for what might be called "theater reconfiguration areas" rather than traditional fixed bases. Such theater reconfiguration areas could be located in remote areas of nations which agree to host them, with a landing strip as the only fixed part of the base. All of the other things needed to prepare equipment and troops for combat could be mobile, concentrating just before an inbound aerial convoy arrived and dispersing as soon as it left. The inventory of supplies at a theater reconfiguration area would be kept to a minimum, and replenished only as necessary. Repair and hospital facilities would also be mobile and dispersed. Theater reconfiguration areas could be protected by conventional concealment methods, electronic masking, and a laser-based missile and air-defense web combining ground-based fire platforms, long-loiter and quick-launch UAV fire platforms, and space-based sensor and fire platforms. Autonomous sentry systems which fall somewhere between a full-fledged robot and a 21st century mobile, smart mine could provide local security. Host-nation support would be kept to a minimum to protect operational security. To complicate targeting by enemies, several decoy theater reconfiguration areas could be set up in each country that allowed them. Such a "shell game" could provide effective deception and thus complicate any attempts to strike at the theater reconfiguration areas with missiles.

A counter-deployment strategy is only one of several asymmetric approaches that future enemies may attempt. They might also resort to terrorism, either in conjunction with a counter-deployment strategy or in lieu of it, to deter American involvement in a regional conflict. In an era when weapons of mass destruction are becoming more common, the terrorism problem is so pressing that some security analysts have begun advocating a retrenchment from global activism is order to lower the chances of provoking terrorism [77]. It may eventually come to that. In lieu of retrenchment, countering an enemy relying on terrorism would require a three part strategy. The first would be to make terrorist attacks more difficult by effective intelligence and by the further hardening of targets. Clearly emerging information technology, including new forms of sensors and new methods for transforming sensor data into usable intelligence, provide part of the solution. The second part would be to institute a policy stating terrorist strikes against the American homeland will provoke a declaration of war against those who use terrorism or sponsor it. Such an approach is a traditional part of war. World War I, after all, began by Serbian sponsorship of terrorism against the Austro-Hungarian Empire. Future sponsors of terrorists— whether the Taliban regime in Afghanistan, Iran, Libya, or some new one—should know that they are performing an act of war and pay accordingly. The third part would be to assure that if the American homeland is struck by terrorism, the result is public support for effective action against the perpetrators rather than disengagement from the conflict that first led to the problem.

Of all forms of asymmetry, urban warfare may be the most problematic and the most likely. In 1996 Ralph Peters wrote, "The future of warfare lies in the streets, sewers, high-rise buildings, industrial parks, and the sprawl of houses, shacks, and shelters that form the broken cities of our world... in the next century, in an uncontrollably urbanizing world, we will not be able to avoid urban deployments short of war and even full-scale city combat" [78]. But as Thomas Ricks notes, urban warfare is one arena where the innovation associated with the revolution in military affairs so far "hasn't helped" [79]. Even the General Accounting Office has noted the inability of the U.S. military to conduct urban operations [80].

Admittedly, few military activities are more difficult than combat in a modern city. Major General Robert H. Scales writes:

> A large urban center is multi-dimensional. Soldiers must contend with subterranean and high-rise threats. Every building could be a nest of fortified enemy positions that would have to be dug out, one by one. Moreover, an experienced enemy could easily create connecting positions between buildings. With limited maneuver space, the urban environment precludes mobility operation and largely negates the effects of weapons, while minimizing engagement ranges. The proximity of buildings plays havoc with communications, further adding to command and control difficulties. Finally, the psychological effects of combat on soldiers are magnified. While the array of threats from multiple dimensions has a debilitating effect on soldiers, it further hastens the disintegration process that haunts all military units locked in close-combat operations [81].

Such fighting involves six key dilemmas: (1) coordination among military units is complicated by separation into small units and by the fact that tall buildings can limit the range of radio signals; (2) it is slow and tedious, nullifying the advantage in maneuver and decisional speed that an advanced military has over less-advanced opponents; (3) it is difficult to distinguish combatants and noncombatants; (4) the battlefield is often thick with noncombatants; (5) holding control of an area is often more difficult than the initial clearing, since enemy troops may reinfiltrate; and (6) since cities are concentrations of communications and information links, operations there will be transparent, broadcast around the world by a variety of means from cell phones to web cams linked to satellite modems. In combination these dilemmas pose an extraordinarily thorny problem.

There are actually three different types of operations that the U.S. military might have to perform in urban areas: policing, raids, and sustained combat [82]. Of these, sustained combat is the hardest. As the 1997 report of the National Defense Panel phrased it, "Urban control—the requirement to control activities in the urban environment—will be difficult enough. Eviction operations—the requirement to root out enemy forces from their urban strongholds—will be even more challenging" [83]. Part of the solution is better doctrine, training, and rules of engagement [84]. The U.S. Marine Corps is far ahead of the other services in this arena. In their *Urban Warrior* experiment, which took place in Oakland, California in 1999, the Marines explored the utility of existing technology like palm-held computers, unmanned aerial vehicles and parachutes steered by the Global Positioning System in an urban battle [85]. At the same time, the Marines are exploring different ways of organizing units involved in urban combat, particularly less hierarchical, more networked structures [86]. During military revolutions, organizational and conceptual change is nearly always more difficult than the adoption of new technology. This certainly holds for urban combat. Joel Garreau notes, "An electronic network may give the Marines unprecedented flexibility, adaptability and competitiveness, but it may also fundamentally unravel the way the Marines have worked for more than 200 years" [87].

Even existing technology is inadequate for urban operations [88]. Two types of technology, though, might help alleviate some of the challenges: nonlethal weapons and robotics. The utility of stand-off, lethal strikes, even if they are substantially more precise than those available today, will remain limited in urban warfare. City fighting involves close combat, often in the presence of noncombatants. Nonlethal capabilities might enable the U.S. military to overcome enemy forces from urban environments with minimal civilian casualties and limited risk to American forces. If nonlethal weapons were developed which could temporarily incapacitate people, separating combatants and noncombatants would entail much less risk to U.S. forces. To hold areas already cleared, nonlethal weapons could limit the risk to U.S. soldiers on sentry duty and lessen the chances that noncombatants wandering through cleared areas would be harmed. For refugee control—which is a vital but often overlooked dimension of urban combat—nonlethals could help stop riots and assist U.S. forces in dealing with any combatants who attempted to hide among refugees [89].

An Army-sponsored workshop at the Jet Propulsion Laboratory, which brought together military professionals and robotics experts, was prescient when it noted that robots hold particular promise for information gathering, the highest priority mission in urban combat operations [90]. Satellites and overhead sensors can never provide the sort of dynamic, three-dimensional picture necessary for urban operations. Robotics have the potential to offer the horizontal perspective to augment overhead sensors. In addition, robotics can form part of a "dynamic perimeter" to guard prisoners and prevent the reinfiltration of cleared areas [91]. The most useful way of penetrating enemy-controlled areas might be through networks of very small but relatively low resolution robotic sensors, with a full intelligence picture developed through data fusion. The utility of robotic systems is almost endless. In armed conflict they could not only perform reconnaissance functions but also serve as mine detectors and sweepers, smoke or other obscurant dispensers, obstacle deployers or breachers, communication relays, target designators, decoys, ambulances, logistics "mules," mobile shields, or offensive strike systems [92].

Scientists also predict that coming decades will bring a biomechanical revolution as engineering devices are blended with organic ones, thus leading to various types of cyborgs. In early 2000, scientists combined a human cell with an electronic circuitry chip. By controlling the chip with a computer, scientists say they can control the activity of the cell. The computer sends electrical impulses to the cell-chip, triggering the cell's membrane pores to open and activating the cell. Scientists hope they can manufacture cell-chips in large numbers and insert them into the body to replace or correct diseased tissues [93]. From a military perspective, such cyborg platforms may be easier to field than purely mechanical robots. For instance, scientists note that it will be several decades before robots the size of cockroaches will have the mobility of cockroaches, but substantial progress has been made in implanting devices in living cockroaches which allows them to be "steered." In future urban warfare, sensory-carrying cockroaches may be maneuvered by soldiers thus providing information dominance.

Broadly speaking, the opening decades of the 21st century will see both symmetric formal war pitting two modern states, and asymmetric formal war pitting a postmodern military against a modern one. There will be reprises of both the Iran-Iraq War and the Gulf War. In the former, the United States may become indirectly involved, providing support of one kind or the other to an ally. As more and more nations acquire nuclear weapons, formal war between them may come to look more like the India-Pakistan war of 1999 than the Iran-Iraq War. Combatants may launch a few limited conventional strikes and perhaps some cyberattacks, but rely primarily on proxy aggression to remain below the threshold of either massive retaliation by their opponent or economic and political pressure from the rest of the world [94]. It remains to be seen whether another postmodern military will emerge to challenge the United States or whether, as American strategic thinking posits, the postmodern U.S. military will always be able to overcome the asymmetric methods used by modern militaries.

INFORMAL WAR

Informal war is armed conflict where at least one of the antagonists is a nonstate entity such as an insurgent army or ethnic militia. It is the descendent of what became known as low intensity conflict in the 1980s. Like today, future informal war will be based on some combination of ethnicity, race, regionalism, economics, personality, and ideology. Often ambitious and unscrupulous leaders will use ethnicity, race, and religion to mobilize support for what is essentially a quest for personal power. The objectives in informal war may be autonomy, separation, outright control of the state, a change of policy, control of resources, or, "justice" as defined by those who use force.

Informal war will grow from the culture of violence which has spread around the world in past decades, flowing from endemic conflict, crime, the drug trade, the proliferation of weapons, and the trivialization of violence through popular culture. In many parts of the world, violence has become routine. Whole generations now see it as normal. To take one example, Debbie Stothard, an expert on refugees who campaigns for democracy in Myanmar, said of the guerrilla groups there:

> These are people who have not had access to a good education and for whom violence is a way of life. It never occurs to them that mounting a siege on a hospital is actually wrong. They have not lived in a world where detaining someone with force is actually unacceptable. It's as though they came from a different planet . . . [95].

This is not an isolated case. In Latin America, the Middle East, South Asia, Central Asia, Sub-Saharan Africa and, to some extent, the inner cities of the United States, a culture of violence has become so pervasive that it is impossible to quell.

In this setting, informal war will remain common, in part because of the declining effectiveness of states. Traditionally, governments could preserve internal order by rewarding regions or groups of society which supported the government, punishing those which did not, and, with wise leadership, preempting conflict and violence through economic development. In a globalized economy, the ability of governments to control and manipulate the economy is diminished, thus taking away one of their prime tools for quelling dissent and rewarding support. In regions where the state was inherently weak, many nations have large areas of territory beyond the control of the government. And, as political, economic, and military factors constrain traditional cross border invasion, proxy aggression has become a more attractive strategic option. Regimes unwilling to suffer the sanctions and opprobrium that results from invading one's neighbors find that supporting the enemies of one's neighbors is often overlooked. This is not likely to change in coming decades. Finally, the combination of globalization and the Cold War have fueled the growth of an international arms market at the same time that the international drug traffic and the coalescence of international criminal networks have provided sources of income for insurgents, terrorists, and militias. With enough money, anyone can equip a powerful military force. With a willingness to use crime, nearly anyone can generate enough money.

Informal war is not only more common than in the past, but also more strategically significant. This is true, in part, because of the rarity of formal war but also because of interconnectedness. What Martin Libicki calls "the globalization of perception"—the ability of people to know what is happening everywhere—means that obscure conflicts can become headline news. There are no backwaters any more. As suffering is broadcast around the world, calls mount for intervention of one sort or the other. Groups engaged in informal war use personal and technological interconnectedness to publicize their cause, building bridges with a web of organizations and institutions. The Zapatista movement in southern Mexico is a model for this process. The Zapatistas, in conjunction with a plethora of left-leaning Latin Americanists and human rights organizations, used of the Internet to build international support with web pages housed on servers at places like the University of California, Swarthmore, and the University of Texas [96]. This electronic coalition-building was so sophisticated that a group of researchers from the RAND Corporation labeled it "social netwar" [97]. Undoubtedly, more organizations will follow this path, blending the expertise of traditional political movements with the cutting-edge advertising and marketing techniques that the information revolution has spawned.

During the Cold War the strategic significance of low intensity conflicts was determined by their potential to spark superpower confrontation or to escalate into wider fighting. Today and in coming decades, strategic significance of informal wars will be determined both by their potential for contagion through refugee flows or terrorism, and by the global image of them which coalesces or is created, whether by participants or other interested parties. A defining feature of the information revolution is that perception matters as much as tangible things. This will certainly hold for informal warfare. Future strategists will find that crafting an "image assessment" or "perception map" of a conflict will be a central part of their planning. While 20th century military strategists like Eisenhower and Marshall took their cues from industrial management, 21st century military strategists must learn from the advertising and marketing industries.

Combat in future informal war is likely to remain "hands on," pitting the combatants in close combat. In many cases, fighting will take place in heavily populated areas. Warriors will be interspersed among noncombatants, using them as shields and bargaining chips. At times, refugee disasters will be deliberately stoked and sustained to attract outside attention and intervention. Informal wars will also be the kind where passion—that most dangerous element of Clausewitz's trinity—plays the greatest role. Unlike formal war, where the trends are toward precision and depersonalization through stand-off capabilities, informal war will remain dirty and bloody, driven by hatred more than science.

In failed states, informal war may be symmetric as militias, brigand bands, and warlord armies fight each other. At other times, it may be asymmetric as state militaries, perhaps with outside assistance, fight against insurgents, militias, brigands, or warlord armies. For the United States, the asymmetric form will be especially important since the American military may be asked to support friendly regimes, contribute to multinational intervention forces, provide humanitarian relief, or even participate in direct combat. This might involve stability operations where U.S. forces, in conjunction with allies, will seek to restore order or facilitate humanitarian relief, and then turn over responsibility for long-term amelioration of the conflict to some other agency or organization. In all probability, multinational mechanisms for the reestablishment of stability and for conflict resolution will grow and improve in coming years. Quite possibly, this will be the major task of the United Nations.

Quick operations to restore stability will be taxing but feasible. Counterinsurgency, which uses military forces to attain not only the short-term restoration of order but also ultimate resolution of the conflict that led to disorder in the first place, is a different and more difficult matter. It involves long-term engagement and alteration of a country's political, economic, security, and even social order. Current American thinking on the security environment and military strategy discounts insurgency and counterinsurgency. Ten years ago they received a moderate amount of thinking in doctrine and strategy: now they are largely ignored. If insurgency is defined solely as rural leftist warfare—its most common and successful variant from the 1940s to the 1990s—then it might make sense to relegate it to history. Maoist "people's war" is unlikely to pose serious problems in the 21st century. But if insurgency is seen more broadly as protracted, asymmetric warfare waged by an organization with a strategic perspective, then the chances are that it will mutate, reemerge and pose challenges to American allies in coming decades. Just as in the 1960s and 1980s, the future U.S. military will have to rediscover counterinsurgency and relearn the lessons of the past [98].

As external sponsors have faded away and state militaries began to understand Maoist people's war, the chances of it working declined. Future insurgents will have to develop new strategies. Every insurgent strategy must have three components: a method for defending the movement against government security forces; a method of raising support; and, a method of attaining ultimate success. In Maoist people's war, insurgent movements defended themselves by tactical dispersion, interspersion among noncombatants, the use of complex terrain such as jungles, mountains, or cities, high internal morale, and effective intelligence and counterintelligence. They supported themselves by using political and psychological means to mobilize internal backers, by taxing citizens and businesses in "liberated" or semi-liberated zones, by capturing arms and supplies from security forces, and by external patronage, whether from a state like the Soviet Union, Cuba, China, and Libya, or a network of ideological allies like a diaspora community (e.g., the Malaysian communists raised money from Chinese communities throughout the Asia-Pacific region, Irish insurgents have used their ethnic brothers in the United States, and so forth). Finally, old-style insurgents sought success by exhausting the government, weakening it through guerrilla war, terrorism, and political warfare, and simply outlasting it.

Future insurgents would need to perform the same functions of defense, support, and the pursuit of victory, but will find new ways to do so. In terms of defense, dispersion is likely to be strategic as well as tactical. There will be few sanctuaries for insurgent headquarters in an era of global linkages, pervasive sensor webs, and standoff weapons, so astute insurgents will spread their command and control apparatus around the world. Information technology will make this feasible. Right wing anti-government theorists in the United States have already developed a concept they call "leaderless resistance" in which disassociated terrorists work toward a common goal and become aware of each other's actions through media publicity [99]. The information revolution will provide the opportunity for "virtual leadership" of insurgencies which do not choose the anarchical path of "leaderless resistance."

Mao Zedong, Ho Chi Minh, Pol Pot, Daniel Ortega, Jonas Savimbi, Fidel Castro and other 20th century insurgent leaders needed physical proximity to their top lieutenants. Twenty-first century insurgent commanders will be able to exert at least a reasonable degree of control from a lap top computer with a satellite modem and web cam situated anywhere in the world, with their transmissions encrypted and bounced throughout the web in order to complicate tracing. The top leadership might never be in the same physical location. The organization itself is likely to be highly decentralized with specialized nodes for key functions like combat operations, terrorism, fund raising, intelligence, and political warfare. In many cases, insurgent networks will themselves be part of a broader global network unified by opposition to the existing political and economic order. For instance, an insurgent network attempting to overthrow the government of a state friendly to the United States might cultivate loose ties with a range of titular allies including global criminal cartels, anti-government groups within the United States, or other political groups seeking to constrain American power.

Unless some sort of new ideological division emerges among the world's great powers—which is not inconceivable—future insurgents will be unlikely to find state sponsors. The trend will be toward "stand alone" insurgent movements that rely on the open market.

Because of this, the revenue-generating node of an insurgent movement will be one of the most important. This commercialization of insurgency has been underway ever since the end of the Cold War cut off ideological patronage. In Colombia, Peru, Kosovo, and other areas, insurgents have found drug trafficking a lucrative source of income. In Sierra Leone and Angola, it is diamond smuggling. But reliance on a single source of income is a vulnerability. Future insurgents may be diversified in their fund-raising methods, using cybercrime as well as traditional methods like extortion, robbery, kidnapping, smuggling, and drug trafficking. They might even move into legitimate commercial ventures, undertake fund-raising among "like thinking" organizations around the world (making heavy use of the Internet), and "tax" co-ethnic diasporas. Money will allow future insurgents to contract out key functions such as fundraising, intelligence and, perhaps, even direct military action. Well-financed insurgents will be able to buy the state-of-the-art talent in key areas like information security or offensive information warfare, thus making them equal or superior to the security forces confronting them. And by contracting out their armed actions, they will lessen the risk to themselves.

Countering new style insurgency will not be easy. There is no formal doctrine for dealing with networked opponents, be they existing criminal cartels or future insurgents. To be successful against future insurgents, the U.S. military will need better intelligence, better force protection, and greater precision at the tactical and strategic levels. In part, these things require new organizational methods. For instance, John Arquilla and David Ronfeldt contend that to match networked opponents, governments must develop network/hierarchy hybrids like those taking shape in the corporate world [100]. The American military also must refine its conceptual tool kit. Ideas like phased operations and centers of gravity, which originated in response to industrial age warfare against hierarchical enemies, will provide little insight into dealing with networked ones.

Emerging technology also holds promise. Again, nonlethal weapons and robotics may prove the most vital. Robotic sensor webs could help with intelligence collection which is always one of the most difficult and most vital aspects of counterinsurgency. With better intelligence, greater precision becomes possible. It might be possible, for instance, to identify and neutralize insurgent leaders with little or no collateral damage or civilian casualties. Removing insurgent leaders does not automatically lead to victory: that requires amelioration of the tensions that opened the way for the insurgency in the first place. But solving root causes is certainly easier with insurgent leaders and cadre out of the way. Nonlethal weapons and robotics also hold great promise for helping to protect any American forces that become involved in counterinsurgency. The lower American casualties, the greater the chances that the United States would stick with a counterinsurgency effort over the long period of time that success demands.

Informal war in the coming decades will not represent a total break with its current variants. It will still entail hands on combat, with noncombatants as pawns and victims. Insurgents, militias, and other organizations which use it will seek ways to raise the costs of conflict for state forces. State forces, whether modern or postmodern, will simultaneously seek ways to impose stability or, in some cases, defeat their opponents at an acceptable cost. It is vital to remember, though, that informal and formal war will be inextricably linked. Interconnectedness and the proliferation of weapons of mass destruction are raising the costs and risks of formal war. States which use traditional force against enemies will often run the risk of retaliation by weapons of mass destruction or, at least, of severe economic pressures from the global financial community which does not look favorably on the market dislocations caused by war. As a result, states will turn more and more to proxy violence through which they might gain their objectives while staying below the threshold which would lead to the use of weapons of mass destruction or to serious economic consequences. The core strategic dilemma for future leaders will be identifying that threshold.

GRAY AREA WAR

As the Cold War ended defense analysts like Max G. Manwaring noted the rising danger from "gray area phenomena" that combined elements of traditional warfighting with those of organized crime [101]. Gray area war is likely to increase in strategic significance in the early decades of the 21st century. To an extent, this is a return to historical normalcy after the abnormality of the Cold War. Militaries have long confronted both "big" and "small" enemies, protecting state territory from foreign invasion while fighting bandits, pirates, and brigands. When foreign invasion was a major concern, armed forces tended to concentrate on it. When it was not, they often spent more of their time and effort on internal order or "small" enemies. This is certainly within the American tradition. Throughout most of U.S. history the Army and, to a lesser extent, the Navy focused on bandits, pirates, and brigands rather than preparing to fight other states in major wars.

Today, gray area threats are increasing in strategic significance. Information technology, with its tendency to disperse information, shift advantages to flexible, networked organizations, and facilitate the creation of alliances or coalitions, has made gray area enemies more dangerous than in the past. For small or weak countries, the challenge is particularly dire. Not only are their security forces and intelligence communities less proficient, but the potential impact of gray area threats is amplified by the need to attract outside capital. In this era of globalization and interconnectedness, prosperity and stability within a state are contingent on capital inflows–except in nations that possess one of the very rare high-payoff natural resources like petroleum, capital inflows require stability and security. In places like Colombia, South Africa, Central Asia, and the Caucuses, foreign investment is diminished by criminal activity and the insecurity it spawns. This makes gray area threats a serious security challenge. It also means that the United States, as the engineer of world order, must take them more seriously.

Gray area war involves an enemy or a network of enemies that seeks primarily profit, but which has political overtones and a substantially greater capability for strategic planning and the conduct of armed conflict than traditional criminal groups. Like future insurgents, future networked gray area enemies may have nodes that are purely political, some political elements that use informal war, and other components that are purely criminal. This greatly complicates the task of security forces that must deal with them. Because gray area enemies fall in between the realm of national security and law enforcement, the security forces that confront them must also be a "gray" blend of the military and the police. Like the military, security forces must have substantial fire power (both traditional or informational), and the ability to approach problems strategically (i.e., to integrate agencies and elements of power, undertake long-term force development, and to think in terms of ultimate objectives and phased programs to attain them). But these security forces also must have characteristics of law enforcement, working within legal procedures and respecting legal rights.

In the opening decades of the 21st century, it will make sense to talk about both *strategic* and *astrategic* gray area war. The strategic form will be that used by some coherent organization or, more likely, network of organizations driving toward a specific purpose. Even though the objective will be monetary rather than purely political, violence will be goal-oriented. Astrategic gray area war will consist primarily of turf battles between armed gangs or militias. It may be related to refugee movements, ethnic conflict, ecological degradation, or struggles for political power (as in Jamaica in the 1990s, where political parties used street gangs to augment their influence). When astrategic gray area war is linked to struggles for political power, the armed forces (such as they are) will be serving as mercenaries only partially controlled by their paymasters, rather than armed units under the actual command of political authorities. Even astrategic gray area war, though, will have security implications since it can deter investment and growth, draw in outside intervention, and, potentially, spark wider armed conflicts. As with many types of future war, the challenge will be the connections and linkages. A single gray area war alone may not be a serious challenge to a major state or a major alliance, but when a number of gray area organizations are linked, or when gray area organizations are connected to other types of threats, the danger will increase.

Since gray area war overlaps and falls in between traditional national security threats and law enforcement issues, states must often scramble to find the appropriate security structure to counter it. Nations with a French administrative tradition have an advantage in that they are comfortable with the idea of a national *gendarmerie* which overlaps military and police functions. As the debate within the United States over the use of the military to counter gray area enemies intensifies in coming years, creation of an American national *gendarmerie* should be considered. Such an organization could combine elements of the military, the intelligence community and law enforcement agencies like the Drug Enforcement Agency and Federal Bureau of Investigation. It could form its own alliances with similar security forces around the world and operate more effectively against gray area enemies in an interconnected security environment and globalized economy.

Gray area war will also pose serious legal and civil rights questions. Should enemies which use it be treated as criminals, with full legal protection, or as military combatants, protected by the law of armed warfare? And, what sorts of legal and ethical frameworks will apply as gray area war spills across borders and becomes increasingly transnational? Even today the United States creates political problems by applying domestics laws on drug trafficking and terrorism to the citizens of other countries, sometimes ignoring normal extradition procedures [102]. This problem is likely to escalate as gray area enemies proliferate and coalesce into networks. The logical response may be an updating of the traditional international law dealing with piracy which gave any nation the right to apprehend and punish a pirate on the high seas. Perhaps this should also apply to future gray area pirates operating in arenas like cyberspace.

The danger from gray area problems should not be underestimated. If left unchecked, gray area conflict can mutate onto informal or even formal war, as one state uses pressure or even force against another which is providing sanctuary to criminals (or, at least, is looking the other way). As a general rule, the lower the level that an armed conflict can be resolved, the less the danger. Concerted effort to thwart gray area war in coming decades can prevent it from becoming even more dangerous.

STRATEGIC INFORMATION WARFARE

Formal, informal, and gray area war are all logical extensions of existing types. Technology, though, could force or allow more radical change in the conduct of armed conflict. For instance, information may become an actual weapon rather than simply a tool that supports traditional kinetic weapons. Future war may see attacks via computer viruses, worms, logic bombs, and Trojan horses rather than bullets, bombs, and missiles. This is simply the latest version of an idea with recent antecedents in military history. Beginning with the writings of people like Guilio Douhet in 1930s, some strategic thinkers held that it might be possible to defeat an enemy state by attacking its homeland directly, bypassing its military forces in the field [103]. Strategic bombing alone did not bring Germany to its knees in World War II (although the theory was more nearly implemented against Japan, which still had a very large proportion of its army intact in the summer of 1945). But for its advocates, this did not disprove their position but simply showed that the technology of the time was immature.

Eventually nuclear weapons did make it possible to destroy a state without fighting a single engagement with its armed forces. But the political utility of nuclear weapons was always subject to question. Their destructiveness was so immense that even a state that waged a "successful" nuclear war would have found it a Pyrrhic victory [104]. By the 1960s, the arsenals of the nuclear powers were extensive enough that it seemed that the only real purposes of these weapons was to deter their use by others and, possibly to deter full-scale invasion of the homeland. Very much the ultimate hammer, nuclear weapons could not be used in instances that called for the modulated use of force.

Proponents of strategic warfare contend that technology now allows their theory to be applied. Information technology might provide a politically usable way to damage an enemy's national or commercial infrastructure badly enough to attain victory without having to first defeat fielded military forces. During World War II, the Germans and Japanese mitigated the effects of strategic bombing by dispersing their productive capacity. The only counter response of the Allies was massive, sustained bombing of every conceivable target. This was inefficient and caused extensive collateral damage (which would now be politically unacceptable). Modern economies are so tightly linked and interdependent that destroying a few key components, particularly communications and power grids, could lead to a cascading collapse of the whole system.

Today strategic information warfare remains simply a concept or theory. The technology to wage it does not exist. Even if it did, strategists cannot be certain strategic information warfare would have the intended psychological effect. Would the destruction of a state's infrastructure truly cause psychological collapse? Would the failure of banking, commercial, and transportation systems crush the will of a people or steel it? After all, everyone who has attempted to use concerted strategic bombing, whether the Germans and the Allies in the World War II or the Americans in Vietnam, underestimated the willpower of their enemies. But until infrastructure warfare is proven ineffective, states and nonstate actors which have the capacity to attempt it probably will, doing so because it appears potentially effective and less risky than other forms of armed conflict.

Future infrastructure war could take two forms. In one version, strategic information attacks would be used to prepare for or support conventional military operations to weaken an enemy's ability to mobilize or deploy force. The second possible form would be "stand alone" strategic information warfare. This might take the form of a sustained campaign designed for decisive victory or, more likely, as a series of raids designed to punish or coerce an enemy. Facing a future Iraq or Serbia, for instance, the United States could conceivably use strategic information attacks rather than aerial bombardment, in part because of the belief that such actions would provoke less political opposition. All of this is, however, speculation. Today the technological feasibility, psychological effect, and legal ramifications of strategic information warfare remain unclear.

But should cyberattacks, whether as part of strategic information warfare or as terrorism, become common, the traditional advantage large and rich states hold in armed conflict might erode. Cyberattacks require much less expensive equipment than traditional ones. The necessary skills exist in the civilian information technology world. One of the things that made nation-states the most effective organizations for waging industrial age war was the expense of troops, equipment and supplies. Conventional industrial-age war was expensive and wasteful. Only organizations that could mobilize large amounts of money, flesh, and material could succeed at it. But if it becomes possible to wage war using a handful of computers with internet connections, a vast array of organizations may choose to join the fray. Nonstate organizations could be as effective as states. Private entities might be able to match state armed forces. Private or commercial organizations might even wage information war on each other—cyber "gang wars" played out on servers and network backbones around the world rather than in ghetto alleys.

As one of the world's most "wired" nations, strategic information warfare could be particularly problematic for the United States, forcing policymakers and military strategists to examine some of their most basic beliefs about warfighting and national security. For instance, the very existence of an infrastructure attack as well as its source could be hidden, at least for a while. An extensive series of problems and system failures induced by an infrastructure attack could occur before the United States understood that it was under attack. It is easy to imagine how tempers would flair if some American defense official in the future had to tell the president that the United States was at war but it was impossible to identify the enemy.

Strategic information warfare would raise a plethora of ethical, political, and legal issues. If the United States was facing a high-tech insurgent, criminal, or terrorist movement, for instance, could the American military (or some other branch of government) strike at its information and financial assets even though they were spread out in computer networks in dozens of sovereign nations? Should cyberattacks be answered only in kind or might traditional weapons be used to respond to them? And, how does the concept of collateral damage apply to cyberattacks? At an even broader level, who is responsible for the defense of a nation's information infrastructure? The government? The military? Private industry?

At the same time that basic policy issues are being discussed, the Clinton administration has begun addressing organizational questions. The first major step was the creation of the President's Commission on Critical Infrastructure Protection [105]. Efforts to protect U.S. information systems have centered on the National Infrastructure Protection Center of the FBI. [106]. This includes representatives of the FBI, the Central Intelligence Agency, the Defense Department, the Secret Service, NASA, and the U.S. Post Office. In addition to assisting with criminal investigations of cyberattacks, it makes information on weaknesses in software available to the public. And, following a number of major denial-of-service hacker attacks on large commercial Internet sites in February 2000, President Clinton announced an initiative to create a voluntary, private-sector network to monitor and respond. Participants will include Charles Wang, chairman of Computer Associates International Inc.; Howard Schmidt, chief information security officer at Microsoft Corp.; Harris Miller, head of the Information Technology Association of America; and "Mudge," a member of a hacker think tank that does security consulting under the name AtStake. The President plans to ask Congress for $9 million to help create the center [107]. According to the White House, the centerpiece of the federal government's efforts in this area will be the Institute for Information Infrastructure Protection, for which the President has requested $50 million in his Fiscal Year 2001 budget [108].

While substantial movement is underway on the defense of national information infrastructure, offensive information warfare is more controversial [109]. Following the 1999 air campaign against Serbia, there were reports that the United States had used offensive information warfare and thus "triggered a superweapon that catapulted the country into a military era that could forever alter the ways of war and the march of history" [110]. According to this story, the U.S. military targeted Serbia's command and control network and telephone system. Other press reports, though, suggested that whatever offensive information warfare capabilities the United States had were not used against Serbia due to ethical and practical problems [111].

Since the cascading effects of information attacks cannot be predicted or controlled given current technology, there were fears that their use would make American military commanders liable to war crimes charges. In January 2000, though, U.S. Air Force General Richard Meyers, then commander of U.S. Space Command, announced that his organization will be given the mission of "computer attack" [112]. The irony is that pressure exists to make the use of force *both* less lethal and more precise. At the end of the 20th century, information warfare is less lethal but also less precise than conventional force. If this changes, strategic information warfare could be catapulted to a central role in U.S. military strategy.

TECHNOLOGICAL TRANSFORMATION

There are glimmerings of changes in war even more profound than strategic information warfare. While they will not alter the essence of war, new technologies or new combinations of technology have the potential to alter not only tactics and operational methods, but military strategy itself by the second or third decade of the 21st century. One of the most important trends in military strategy between the 18th and 20th centuries was the broadening of its focus. In the 18th century, one needed only to destroy the enemy's field army or, in some cases, seize control of key forts or territory. With the emergence of "total war" in the 20th century, an enemy's entire society and infrastructure became the targets of military operations. Modern technology allowed war to move toward a "total" form described by Clausewitz, reaching ever greater levels of destruction. The conundrum faced by political leaders today is that there is still a need to use armed force, but interconnectedness and other factors have made it difficult to mobilize and sustain the level of passion and hate necessary for total war. Strategists thus need some way to coerce or punish an enemy elite or, at least, to disrupt their plans, without the wholesale destruction of infrastructure or killing of noncombatants. This is the reason that precision is such an integral element of the current revolution in military affairs.

Within a few decades, technology may provide solutions to this strategic conundrum. After the Gulf War, American military leaders bragged that technology allowed them to not only select which building a bomb will hit, but to select which window of the building the bomb entered. Soon technology, particularly mini- or micro-robots, may allow military planners to select which individual or physical object in a building is to be destroyed. For the first time, it might be possible to target only the aggressor's leaders, leaving noncombatants untouched. Within a few decades the technology might exist to construct killer robots the size of a grain of sand that could search for and kill future Saddam Husseins.

Like all new military technology, such fine-tuned precision will bring new risks, costs, dilemmas, and unintended side effects. Americans have long struggled with the ethics of deliberate assassination of enemy leaders. Such acts were rare even in the midst of declared war. During World War II, the only known instances were the American downing of the plane carrying Admiral Yamamoto, a British attempt to kill Erwin Rommel, and a German plot to kill Dwight Eisenhower. Today, assassination of enemy leaders outside of declared war is proscribed by presidential directive. But as the technology to target enemy elites becomes available, Americans (and any others who develop a postmodern military) may rethink the ethics of using it. Future armed conflict may no longer pit one society against another, but one leadership cadre against another [113]. While much speculation on future war focuses on the proliferation of weapons of mass destruction and the spread of terrorism and thus contends that noncombatants will be prime victims of future wars, the opposite is at least feasible. With brilliants robots, future armed conflict, like much of medieval war and 18th century European war, may be a sport for elites that leaves the masses relatively untouched.

This is only the tip of the technological iceberg. Coming decades are likely to see the proliferation of robots around the world and in many walks of life. Hans Moravec, for instance, contends that mass produced robots will appear in the next decade and slowly evolve into general purpose machines [114]. Ray Kurzweil takes the argument even further and holds that by the end of the 21st century, human beings will no longer be the most intelligent entities on the planet [115]. However fast the evolution of robotics proceeds, it will invariably affect armed conflict. As one of the most avid customers of new technology, this will certainly affect the American military in the years after 2020.

Initially, the prime function of military robots will be to replace humans in particularly dangerous or tedious functions. Examples might include evacuation of casualties under fire; operating in environments where nuclear, biological, or chemical weapons have been used; mine clearing; firefighting; and reconnaissance, surveillance, and target acquisition [116]. The real breakthrough and decision point will come when robots advance to the point that they have the potential for combat use. This will take some time, particularly for land warfare which takes place in a much more challenging operating environment for autonomous systems than does air, space, or sea warfare. Robots intended for battlefield use will have to be orders of magnitude "smarter" than those used for less stressful functions such as loading and moving material [117].

Current thinking about the technological characteristics of future military robots moves along two parallel tracks, each synthesizing robotics and other emerging technologies. The first envisions autonomous systems that employ sensors, computing, and propulsion very different from that used by people. One of the goals in this arena is miniaturization. Mini or micro-robots could be easily carried, yet perform a range of difficult or dangerous military missions [118]. The Pentagon already has a $35 million program under way to develop a bird-like, flapping-wing micro-air vehicle for battlefield reconnaissance and target acquisition [119]. But this is just the beginning, the true revolution could come from the maturation of microelectromechanical systems or MEMS which many leading scientists contend will be developed within 30 years with a dramatic impact in many endeavors.

MEMS technologies construct very tiny mechanical devices coupled to electrical sensors and actuators. [120]. According to the Defense Advanced Research Projects Agency (DARPA):

The field of Microelectromechanical Systems (MEMS) is a revolutionary, enabling technology. It will merge the functions of compute, communicate and power together with sense, actuate and control to change completely the way people and machines interact with the physical world. Using an ever-expanding set of fabrication processes and materials, MEMS will provide the advantages of small size, low power, low mass, low-cost and high-functionality to integrated electromechanical systems both on the micro as well on the macro scale.

MEMS is based on a manufacturing technology that has had roots in microelectronics, but MEMS will go beyond this initial set of processes as MEMS becomes more intimately integrated into macro devices and systems. MEMS will be successful in all applications where size, weight and power must decrease simultaneously with functionality increases, and all while done under extreme cost pressure [121].

Eventually MEMS could open the way for an even more profound revolution in nanotechnology which is based on "bio-mimicry" manufacturing. A report from the U.S. Commission on National Security/ 21st Century states:

The implications of nanotechnology are particularly revolutionary given that such technologies will operate at the intersection of information technologies and biotechnologies. This merging and melding of technologies will produce smaller, more stable, and cheaper circuitry that can be embedded, and *functionally interconnected*, into practically anything—including organic life forms [122].

In the military realm, MEMS and nanotechnology could allow things like a "robotic tick" the size of a large insect which could attach itself to an enemy system such as a tank, then gather and transmit information or perform sabotage at a designated time [123]. In a fanciful but technologically feasible description of the future battlefield, James Adams writes:

MEMS opens a window on a new generation of technology that will literally transform the battlefield. Tomorrow's soldier will go to war with tiny aircraft in his backpack that he will be able to fly ahead of him to smell, see and hear what lies over the hill or inside the next building.

Additional intelligence will be supplied by sensors disguised as blades of grass, pockets of sand or even clouds of dust [124].

However radical such a notion might seem, it is, like the official vision of the future, essentially new technology used in old ways. By contrast, futurists like Martin Libicki have speculated on modes of warfare to make maximum use of MEMS-based technology. In fact, Libicki's alternative vision of future war is one of the most profound and creative seen to date. Its essence is that information technology, among other things, is shifting the advantage in warfare to "the small and the many" over "the large, the complex, and the few." This is in stark contrast to orthodox American strategic thinking that seeks ever more capable systems that are, by definition, more expensive, and thus acquired in smaller numbers, but is a logical development of the concept of distributed robotics under exploration by DARPA.

Libicki describes three stage in the ascendance of "the small and the many." He calls the first "popup warfare." This is based on extant technology in a security environment characterized by the proliferation of precision guided munitions (PGMs). While *Joint Vision 2010* and other official documents expect many states to have precision guided munitions, they assume that the American military can overcome enemy PGMs by stealth, operational dispersion, and speed. Libicki is more skeptical. "The contest between stealth and anti-stealth will be long and drawn-out," he writes, "but...the betting has to be against stealth for any platform large enough to encompass a human...even with stealth, everything ultimately can be found" [125]. The result will be "popup warfare" where both sides stay hidden most of the time, pop up just briefly to move or shoot, and then "scurry into the background" [126].

Libicki's second stage of future warfare, which he calls "the mesh," uses technologies available over the next 20 years against an enemy with developed industry but underdeveloped informational capabilities. To a large extent, this is coterminous with the official vision that calls for an interlinked mesh of sensors and information technology to give American commanders a clear and perfect view of the battlefield while their opponents remain in the dark. Reinforcing the assumptions found in *Joint Vision 2010* and other official documents, Libicki writes, "Tomorrow's meshes will allow their possessor to find anything worth hitting" [127].

Libicki's third stage represents the ultimate ascendance of "the small and the many." He contends that eventually enemies will develop their capabilities to the point that the platforms that compose the American military's "mesh" will be vulnerable to attack. The solution is to weave a mesh composed of small, moderately priced objects rather than a handful of very large and very expensive ones. "Battlefield meshes, as such, can be built from millions of sensors, emitters, and sub-nodes dedicated to the task of collecting every interesting signature and assessing its value and location for targeting purposes" [128]. This is where MEMS-based robotics becomes significant. Libicki speculates on the value of ant-like robots with each one having a fairly limited capability, but the weaving together of their collective capabilities generates extensive capabilities. The inherent redundancy of the mesh in what Libicki calls "fire ant warfare," in which small, relatively simple weapons and sensors swarm onto a large complex one as a means of attack, would make it much more robust than the one envisioned in official documents.

While initial thinking about robotics concentrates on miniaturization and the integration of networks of small robots with relatively limited functions, partially organic robots may prove nearly as useful. According to a recent report from the U.S. Commission on National Security/21st Century, "Notions of 'androids,' cyborgs,' and 'bionic' men and women have dwelled exclusively in the realm of science fiction. But at least the beginnings of such capabilities could literally exist within the lifetime of today's elementary school children" [129]. Soon it might be possible to mount cameras or other sensors on dogs, rats, insects or birds and to steer them using some sort of implant [130].

Simple cyborgs like this may be only the beginning of an even more fundamental revolution or, more precisely, the marriage of several ongoing technological revolutions. Lonnie D. Henley, for instance, argues that a melding of developments in molecular biology, nanotechnology, and information technology will stoke a second generation revolution in military affairs [131]. Nanotechnology is a manufacturing process that builds at the atomic level [132]. It is in very early stages, but holds the real possibility of machines that are extremely small, perhaps even microscopic. Eric Drexler, the most fervent advocate of nanotechnology, predicts that it will unleash a transformation of society as self-replicating nanorobots manufacture any materials permitted by the laws of nature and thus help cure illness, eliminate poverty, and end pollution [133]. As Henley points out, combining nanotechnology with molecular biology and advances in information technology could, conceivably, lead to things like biological warfare weapons that are selective in targets and are triggered only by specific signals or circumstances. It could also lead to radically decentralized sensor nets, perhaps composed of millions of microscopic airborne sensors or, at least, a mesh of very small robots as envisioned by Libicki. And, Henley contends, it might eventually be possible to incorporate living neuron networks into silicone-based computers, thus greatly augmenting their "intelligence."

Beyond technological obstacles, the potential for effective battlefield robots raises a whole series of strategic, operational, and ethical issues, particularly when or if robots change from being lifters to killers. The idea of a killing system without direct human control is frightening. Because of this, developing the "rules of engagement" for robotic warfare is likely to be extraordinarily contentious. How much autonomy should robots have to engage targets? As a robot discovers a target and makes the "decision" to engage it, what should the role of humans be? Would a human have to give the killer robot final approval to shoot? How would the deployment of battlefield robots affect the ability of the U.S. military to operate in coalition with allies who do not have them (given that a roboticized force is likely to take much lower casualties than a non-roboticized one)? Should the United States attempt to control the proliferation of military robotic technology? Is this feasible since most of the evolution of robotic technology, like information technology in general, will take place in the private sector? Should a fully roboticized force be the ultimate objective?

Other emerging technologies could prove equally revolutionary. One example is what can be called "psychotechnology." Future military commanders might have the technology to alter the beliefs, perceptions, and feelings of enemies. This could range from things like "morphing" an enemy leader and creating a television broadcast in which he surrenders to much more frightening ideas like perception-altering implants, chemicals, or beams of some sort. Such technologies would be particularly ominous from an ethical perspective. Today, effective and controllable psycho-technology is science fiction, but so too was space flight not so long ago. Any developments in this realm warrant very close scrutiny. Barring some sort of truly fundamental change in the global security environment, they should be eschewed.

PART III: THE MARK OF SUCCESS FOR FUTURE MILITARIES

FOUNDATION

The more things change, the more they stay the same. Like all clichés, this one has a core of truth, particularly for of armed conflict. Colin Gray reminds us, more about strategy is persistent, even eternal, than is changeable [134]. Even in revolutionary times, continuity outweighs change. This holds true for the current revolution in military affairs. No matter how much technology, operational methods, or military organizations shift (or appear to shift), war will always involve, as Clausewitz noted, a dangerous and dynamic relationship among passion, hatred, reason, chance and probability [135]. The best that military commanders and strategists can hope for is to hold passion in check with rationality, and to minimize the deleterious effects of chance and probability through planning, training, and, to an extent, technological aids. One can only prepare for chance through redundancy, fallback positions, and "Plan B." Passion injects even more conundrum, particularly in informal war where its depths can be difficult for those not intimate with the conflict. For some reason, humans find it easier to accept simple bad luck than blind hatred.

The "specialness" of warfighting and warriors will survive any real or apparent changes in the nature of armed conflict. War is and will be distinct from other types of human activity. Largely because of this, future warriors, at least in democracies like the United States, will continue to be bound by an ethos stressing duty, honor, sacrifice, and the highest ethical standards. Sometimes their enemies will not share this. This ethical asymmetry is likely to be significant. In future armed conflict as in past warfighting, quality will continue to trump quantity to a large extent. What will change, though, is the definition of quality. It will expand and take on new, important meanings. But this is not surprising: the expansion of concepts and definitions is perhaps the most pervasive change of all in the current revolution in military affairs.

SPEED

One of the most important determinants of success for 21st century militaries will be the extent to which they are faster than their opponents. Increased speed is a defining element of the current revolution in military affairs (and of the information revolution and the revolution in business affairs, for that matter). Armed forces that can move and make decisions rapidly will have inherent advantages. Tactical and operational speed comes from information technology—the "digitized" force—and appropriate doctrine and training. It allows a military to surprise its enemy, to maximize its own advantages and minimize those of the opponent, and to remain dispersed as long as possible (which will be vital in the lethal battlefields of the 21st century thick with precision guided munitions and weapons of mass destruction). A military with tactical and operational speed will be able to "pulse" its activities, concentrating and dispersing in rapid sequence. Just as there are specific frequencies at which a pulsed strobe light is most unsettling to those exposed to it, it is likely that there are psychologically optimal frequencies for pulsed military operations.

Strategic speed will be equally important as a determinant of success in future armed conflict. For nations that undertake long-range power projection, strategic speed includes mobility into and within a theater of military operations. In the broadest sense, this entails making a lighter, more transportable military force. A number of technologies show promise in this arena, from relatively simple steps like replacing heavy materials in military equipment with lighter ones—from steel to carbon fiber or woven metal, for instance—to more complex ones like eventually moving away from petroleum based fuels and kinetic ammunition, both of which are heavy and bulky. Another step that could help with strategic speed is an emphasis on modularity in all military systems. Rather than having a tank that could only be a tank, a truck that could only be a truck, etc., it would be very valuable to have a standard mobility platform which could have task-specific modules attached to it.

Strategic speed also entails faster decision-making. One of the core dilemmas the United States is likely to face is having a military that can deploy and operate at lightning speed, while strategic and political decision-making remains a time-consuming process of consensus building. The U.S. Army is currently using the concept of "strategic preclusion" as a tool to assess long-term force development [136]. This calls for an Army that can be deployed to a theater to forestall imminent aggression rather than having to dislodge an aggressor that has already seized the territory of American friend or ally. But this concept assumes that a future American president will use military force very early in a crisis before other options have been exhausted. The American political decision-making process simply does not work that quickly. Consensus building, particularly when it involves the use of force, takes time. This problem is even worse for decision-making in alliances like NATO where each participant must undertake national decision-making prior to collective action. This is not always bad thing. A consensus decision is often better than an imposed one, particularly in any a system where the rank and file have both some degree of power and access to information. Protraction, though, brings risk.

In coming decades, interconnectedness plus the dispersion of power and information will make the need for consensus decision-making even more important. States will have to justify the use of armed force. The problem is balancing the time consuming nature of consensus decision-making with the speed of military operations. The ability to launch quick, perhaps even preemptive military operations will be one of the advantages which postmodern militaries will have over less advanced adversaries. But the states which have the potential to build postmodern militaries will be precisely those with the greatest need for consensus decision-making. This will be an enduring, perhaps even defining dilemma for advanced states. While decision-making speed may not be decisive in future armed conflict, it will be a factor, with some advantages going to those entities that can reach a decision quickly. This suggests an advantage of sorts for postmodern nonstate actors, which often will have more streamlined means for decision-making. Often armed conflict between a postmodern state military and a postmodern nonstate actor will be one where the aggregate resources of the state are used to counterbalance the flexibility and adaptability of the nonstate actors. In such cases, the ultimate outcome will often be contingent on the willingness of the state to bear the costs of the conflict.

Speed also has an even broader, "meta-strategic" meaning. The militaries which meet with the greatest success in future armed conflict will be those which can undertake rapid organizational and conceptual adaptation. In existing militaries, with their bureaucratic and hierarchical organizations, change is slow, often glacial. It can take decades to develop and field new systems, concepts, or methods. Some of the enemies of the future, particularly networked opponents using informal and gray area war, will adapt rapidly. In some cases, this may compensate for quantitative shortcomings. To respond, successful state militaries must institutionalize procedures for what might be called "strategic entrepreneurship"—the ability to rapidly identify and understand significant changes in the strategic environment and form appropriate organizations and concepts. Information technology will help. By allowing intricate, cross-cutting, and broad band communication across components of the military, information technology will speed up the adaptation process. But this is not enough. The institutional cultures of successful militaries will shift toward greater creativity and flexibility and away from a debilitating degree of risk aversion. This will have far-ranging effects on recruitment, leader development, promotion, and education. Military schools like staff and war colleges will be particularly vital. As in the past, victory or defeat in future armed conflict will be rooted in events that took place years earlier in military classrooms.

PRECISION

It is often said that the American military of the future must be "more lethal" [137]. But lethality is only part of future success. In fact, the future American military must be able to operate with greater *precision*. In one sense, this is not a new idea. Most analysis of the current revolution in military affairs stresses precision. George and Meredith Friedman, for instance, rank the development of precision guided munitions along with the introduction of firearms, the phalanx, and the chariot as "a defining moment in human history" [138]. But despite the attention given to precision, the architects and analysts of the revolution in military affairs have taken too narrow an approach to it. Like speed, precision has multiple facets and dimensions.

So far thinking on the revolution in military affairs has focused on what might be called *physical* precision—the ability to hit targets with great accuracy from great distances with precisely the desired physical effect. Physical precision is derived from improved intelligence, guidance systems and, increasingly, from the ability to adjust or "tune" the effects that a particular weapon has. A proposed electro-magnetic gun, for instance, could be adjusted from a non-lethal setting to an extremely lethal one [139]. But there is more to precision than simply hitting the right target. Military strategists and commanders must come to think in terms of *psychological* precision as well.

Psychological precision means shaping a military operation so as to attain the desired attitudes, beliefs, and perceptions on the part of both the enemy and other observers, whether noncombatants in the area of operations or global audiences. Like so much of the revolution in military affairs, this is a new variant of an old idea. Military thinkers have long understood that war is integrally, perhaps even essentially psychological. Sun Tzu, of course, crafted the quintessentially psychological approach to strategy, contending that "all warfare is based on deception" [140]. While some disciples of Clausewitz, particularly German military strategists, acted with disregard for the psychological dimension of strategy, the Prussian theorist himself clearly understood that war was a psychological struggle and the objective is to break the enemy's will [141].

Today the American military is not as strong at psychological precision as it should be, in part because technological advantages appear to make psychological effectiveness unnecessary. The explanation, though, runs even deeper than that. For a nation composed of many cultures, the United States has never had a deep understanding of other cultures, perhaps because it was never a major colonial power. This has shown up whenever the U.S. military is engaged in cross-cultural conflict. Often American strategists "mirror image" the enemy and build their campaigns based on what they feel would cause Americans to surrender without taking into account the psychological differences between antagonists. When the enemy does not react to conditions the way that Americans would, he is labeled "irrational." The American experience with counterinsurgency offers many illustrations: astute thinkers like Edward Lansdale and John Paul Vann who understood it were often ignored. But just as modern corporations are finding that they can no longer afford the inefficiencies of their industrial predecessors, postmodern militaries must extract every possible degree of precision, psychological and well as physical. In fact, so much of 21st century armed conflict will be cross-cultural and played out in the full glare of global scrutiny that psychological precision might be the more important variant.

How might future militaries attain greater psychological precision? To some extent, technology can help. It is vital to have a very wide range of military capabilities—a "rheostatic" capability—to assure that an operation has the desired psychological affect. This suggests a growing need for effective nonlethal weapons, particularly in instances where the psychological objective is to demonstrate the futility of opposition without killing so many of the enemy or noncombatants that the enemy's will is steeled rather than broken or that public opposition is mobilized. Some advocates of nonlethal weapons go so far as to see them as the central element in future armed conflict [142]. This is probably an overstatement; they will be an integral tool for attaining psychological precision and sustaining the political utility of force.

Different forms of psychotechnology might allow greater psychological precision. Conceivably, technology might be developed that would give militaries the ability to alter the perceptions of targets, perhaps causing intense fear, calm, or whatever reaction was required. But any state with the capability and inclination to develop such technology should be extraordinarily careful because of the potential for violations of basic human rights. In the vast majority of cases, technology for psychological manipulation should be eschewed. Some state or organization unbound by ethical and legal constraints, though, eventually may field psychotechnology. Then the United States will have to decide whether to respond in kind or seek other means of defense. The potential for a psychotechnology arms race is real.

Technology, though, is only part of psychological precision. There is a vast body of psychological analysis, particularly that dealing with anxiety and fear, which is not adequately integrated into military planning. When the goal is to create fear and anxiety or collapse the will of an enemy, the operation should be phased and shaped for maximum psychological impact. Successful militaries must take steps to assure that operational and strategic planning staffs are psychologically astute, whether by educating the planners themselves or using information technology to provide access to psychologists, cultural psychologists, and members of other cultures. They should undertake cross cultural psychological studies aimed at building data bases and models which can help guide operational planning.

Precision has a strategic component which is sometimes overlooked. Strategic precision entails shaping a military so that it best reflects its nation's strategic situation, including strategic culture, level of technological development, and most significant threats. For the U.S. military, this entails finding the appropriate balance among capabilities to deal with formal war, informal war, and gray area war. It also entails reaching a degree of privatization which maximizes efficiency without creating unacceptable risks. In attaining strategic precision, past success can be a hindrance. As Edward Luttwak points out, the paradoxical logic of strategy often makes victory the midwife of defeat. Militaries which have won great victories resist change, even when the methods and structures that brought them success grow obsolete [143]. Victory limits the urge to adapt and innovate. For the United States, avoiding a victory-induced slumber will be a key step toward a postmodern military.

Strategic precision will also entail the ability to identify the key strategic thresholds. Nonlethal weapons and strategic information warfare will cause the threshold for the acceptable use of force to be redrawn. In the old security system, states and even nonstate actors knew when force was considered appropriate. At times they deemed the costs and risks of ignoring this worthwhile, but usually formed their strategies with the threshold in sight. But nonlethal weapons and strategic information warfare will increasingly blur this threshold. It will be some time before strategists can reestablish it. Further up the scale, the proliferation of weapons of mass destruction and globalization are elevating the threshold at which states will be willing to undertake the costs and risks of large-scale conventional war, leading them toward less provocative (and less effective) methods such as limited strikes, proxy violence, or information warfare. Eventually, the maturation of miniaturized robots may force an additional redefinition of thresholds as states that have such technology decide when and how it should be used. States that develop an accurate understanding of these new thresholds—those that understand the ethical contours of their time—will be less likely to make dangerous miscalculations, and thus more successful than their less astute counterparts.

FINDING AND HIDING

In an age of precision guided munitions, what can be found usually can be destroyed. As Martin Libicki notes, conventional warfare is changing from force on force to hide-and-seek [144]. Given this, one of the most crucial dynamics of future armed conflict will be the struggle between finding and hiding. Successful militaries will be those better at finding their enemies than their enemies are at finding them. Within the United States, the emphasis has been on the offensive part of this equation—the finding. American strategists and technologists are expending great effort and treasure to build ever more effective systems of systems linking a multiplicity of sensors, developing means for rapid data fusion, and communicating the derived knowledge to battlefield commanders. This is certainly worthwhile: the ability to "find" will be a determinant of success in future armed conflict. Hiding, though, warrants more attention.

Hiding has not been ignored, particularly at the tactical level. Witness all of the effort given stealthy technology in the United States. Campaign plans usually include an appendix outlining the steps for deception and operational security. But the information revolution and interconnectedness have altered some of the basic precepts of deception. Standard works on military deception written as recently as 10 years ago are virtually obsolete [145]. Future military strategists must rebuild their understanding of deception and hiding, working with new information technology that allows morphing and sophisticated spoofing (including things like holographic soldiers, tanks, planes, and so forth). In particular, the notions of operational and strategic deception must be revisited. At the same time, the legal and ethical dimensions of deception need refinement. Militaries which do this before they enter armed conflict will increase their chances of success. Those which do it "on the fly" will face problems.

REORGANIZING

A revolution in military affairs requires not only new technology, but new operational concepts and organizations [146]. The most successful future militaries will be those that undertake a "blank slate" reevaluation of their basic concepts and organizational precepts. Organizations that represent a hybrid blend of hierarchical structures with networks, public components with private, and humans with machines await further analysis. We can borrow from the business world, but not import directly. We know that hybrid will be important, but do not yet know their shortcomings and hidden problems.

Other blank slate organizational questions also need asked. For instance, does it make sense to think of military service as a career that begins at age 18 or 21, continues for 20 or 30 years, and then stops? The dynamics of the information revolution may, in fact, force postmodern militaries to consider things like mid-career accessions in addition to contracting. Since many of the skills needed by future militaries will also be in great demand in the civilian sector, militaries might find it necessary to recruit mid-level or upper level leaders from the private sector directly into the service, and to reward them at a level equivalent to what they would attain in the civilian sector. This is, of course, a risky procedure. It could cause tension and rifts within a military between those who were career military professionals and those who came in mid-career. Steps would have to be taken to assure that that those who joined the military mid-career reflected the ethos of honor, duty, and sacrifice that is essential to the functioning of a military. Some analysts are already voicing concern over the ethical repercussions of privatizing many military functions [147]. Even so, reevaluating career paths in the military might be necessary.

The trend in the commercial world has been toward a blurring between management and staff. If this is extrapolated to the military, it might be necessary to consider whether the division of a service into enlisted personnel and commissioned officers makes sense in the 21st century. After all, this distinction arose to reflect the schism between commoners and aristocrats during the birth of modern militaries. Since societies are no longer organized that way, perhaps militaries should abandon the split between aristocrats and commoners.

In addition, the organization of militaries into land, sea, and air services needs assessed. Perhaps it would make more sense to organize them into components focused on a specific type of armed conflict—one for formal war, one for informal, and one for gray area war. Alternatively, postmodern militaries must consider whether a new service is needed for new operating environments. Martin Libicki, for instance, supports the creation of an "information corps" within the U.S. military [148]. There are reports that China is considering formation of a fourth military service to concentrate on information warfare [149]. Other analysts go even further than calling for a new service. For Robert Bunker, who is one of the more creative writers on future warfare today, this implies a fundamental change in the nature of warfare—not simply the appliqué of micro-processor based technology as in the official vision of the future, but the addition of a fifth dimension of warfare to three-dimensional space and time [150]. In stark contrast to the official vision of the future, Bunker holds that the United States is unlikely to attain dominant battlespace knowledge in cyberspace, whether in what he calls the "upper tier" which is the Internet and the electromagnetic spectrum, or the "lower tier" which is the stealth masking of physical forces. But, he predicts, other state militaries will do no better, so the prime enemy will be non-state actors, often criminals, with the flexibility and creativity to make use of cyberspace's potential. For the U.S. military to be truly successful, Bunker argues, it must master new concepts like cyber-shielding, cyber-maneuver, and what he calls "bond-relationship" targeting that creates "tailored disruption within a thing, between it and other things, or between it and its environment by degrading, severing, or altering the bonds and relationships which define its existence." Whether Libicki's more modest proposal or Bunker's radical one proves accurate for postmodern militaries, it is clear that those which are able to let go of old organizational patterns and embrace, even master new ones will be the most likely to succeed in future armed conflict.

ADJUSTING CIVIL-MILITARY RELATIONS

The ability of a state military to succeed at armed conflict is determined, in part, by its relationship to the society it defends. Stable, healthy civil-military relations make it easier to sustain support for a military and for the military to recruit talented members. At the beginning of the 21st century, the changing nature of armed conflict will force every state military to evaluate and adjust its relationship with society. For instance, the ability of postmodern militaries to strike targets around the world very quickly and with apparent impunity will aggravate the tension between military commanders and civilian leaders [151]. Military commanders will recognize that speed and preemption increase the chances of operational success and decrease risk, while civilian leaders—at least those in democracies—will continue to see armed force as a last resort, only to be used when other methods have failed (and the risks to the military have thus escalated). Eventually the emergence of strategic information warfare and the increased significance of gray area war and infrastructure attacks will blur the line between military and nonmilitary functions.

Since so many factors shape civil-military relations— historical, economic, political, cultural, demographic, and so forth—every military will face a slightly different set of problems as it adjusts. Despite some recent worries about a "crisis" in American civil-military relations, the U.S. military continues to be held in high esteem by the public [152]. This is due, in part, to the great efforts by the leaders of all the services to inculcate the highest ethical standards possible, and to stress sacrifice, honor, and duty. It is also due to the afterglow arising from of the collapse of the Soviet Union and the American-engineered victory in the Gulf War. Eventually this will fade. If the future U.S. military is involved in murky and morally confusing conflicts against nonstate actors and criminal cartels rather than aggressive dictators, public support might be shakier.

In 1992 Charles J. Dunlap, Jr. held that the use of the military in nontraditional roles, especially law enforcement functions, could shake the foundation of civil-military relations [153]. While Dunlap's literary device—a failed future military coup d'état in the United States—was deliberately far-fetched, his point merits serious consideration. The current health of American civil-military relations is based on the perception that: (1) the military has a vital job to do in defending the nation against external enemies, it does so very competently, and should receive adequate resources to do so; (2) the effectiveness of the U.S. military does not threaten domestic civil rights or political institutions; and, (3) the U.S.

military represents the best of traditional American values. Changes in any of these three components could degrade civil-military relations.

This is not to suggest that American policymakers should eschew the use of the military for anything other than traditional warfighting. If informal or gray area war poses serious threats to American national security, the American people and their elected leaders are likely to demand the involvement of the U.S. military. But policymakers must be aware of the potential danger to civil-military relations. Among other things, any decline in the prestige of the military could complicate recruitment. The integration of more and more technology into increasingly rapid and complex operations makes it vital for the military to be able to attract the highest quality recruits. The prestige of service is a vital component of this. If the prestige erodes, it will be very difficult to attract talent, particularly in those realms like information technology and strategic leadership where the military competes directly with the commercial world for talent. Along these same lines, the U.S. military must continue its efforts to assure that its ethnic and gender composition reflects American society (even while insisting that its values reflect the *best* of American society rather than the norm).

The U.S. military must do its part to help forestall problems with civil-military relations. Foremost, it must assure that any capabilities or methods it develops reflect national values and strategic culture. For instance, it should eschew operational concepts that call for the preemptive use of force on the part of the United States or for actions that would indiscriminately harm noncombatants like attacks against satellites or information systems which are linked to global networks. It might be tempting to use hacker attacks to seize the economic or informational resources of, say, a gray area enemy, but if this entails intrusion into the information infrastructure of other states, its adverse political results could outweigh its military utility. Given this, the U.S. military should not even suggest such actions to political leaders. And, unless circumstances change in some fundamental way, the military should eschew development of dangerous new technologies like psychotechnology which run counter to American values like personal privacy and civil rights.

After Vietnam, the U.S. military stopped taking civil-military relations for granted and recognized that it was like a marriage or any other relationship: maintenance required hard work. As a result, the American military established a number of outreach and communications programs to solidify its links with the public and with elected leaders. In the 21st century security environment, successful militaries will emulate this. Those which leave civil-military relations to chance will suffer; those which nurture the relationship through the transition in the nature of armed conflict will have an easier time.

CONTROLLING FOR ASYMMETRY

Since asymmetric conflict will be common in the opening decades of the 21st century, finding ways to resist or transcend it will be one of the determinants of success for militaries and other organizations that participate in armed conflict. When a postmodern state military is pitted against a modern state military or a modern nonstate actor, time will be the key element. Postmodern militaries will attempt to use speed and knowledge to bring the conflict to quick resolution. Their enemies will seek protracted wars, whether through dispersion, deception, terrorism, counter deployment operations, or persistence. This means that postmodern militaries must seek speedy resolution of conflicts. But what will happen if they fail?

Current American thinking about future war is based on the idea that if war becomes necessary, the preferred method is a quick resolution using cutting-edge, rapidly deployable forces and precision strikes against key targets. If sustained combat becomes necessary, then reserve component units (and, hopefully, allied forces) will be mobilized and deployed. Beyond existing reserve component units, plans for the creation, training, and equipping of new units are underdeveloped. Similarly, there are no plans for reconstituting the defense industrial base. Long wars are simply considered inconceivable. This is a potential problem. While everything suggests that the future United States (just like the current one) would prefer short wars, failing to plan for protracted conflict increases the chances it will occur. Given this, greater attention should be given to protracted war in the various wargames, seminars, and simulations that the U.S. military uses to think about future armed conflict.

Armed conflict involving a postmodern nonstate actor is more difficult to assess, in large part because there are no full-blown examples to consider. It is not clear, for instance, whether a postmodern nonstate actor could undertake a protracted conflict or not. In all likelihood, it could since it would not be constrained by the interconnectedness and legal/ethical frameworks that make protracted war difficult for nations. Asymmetric conflict with a postmodern nonstate actor is thus likely to devolve into a contest of wills, with the side willing to pay the greatest price in blood and treasure coming out ahead. Successful 21st century militaries will be those who understand asymmetry, transcend it when possible, and moderate its effects when they cannot transcend it.

ADAPTING TO TECHNOLOGICAL SHIFTS

The ability to accept and capitalize on emerging technology will be a determinant of success in future armed conflict. This is good news for the United States. No military is better at this than the American, in large part because no culture is better at it than the American. Americans are infatuated with technology. This has deep roots in history. As the United States grew and matured throughout the 19th century, the rapid expansion of the frontier led to persistent labor shortages. Technology, by substituting machinery for human muscle, offered a partial solution. What began, then, as a practical reaction to an economic problem eventually had a profound impact on national perceptions and attitudes, settling deep within the American collective self-conscious. It is more than coincidence that much of modern, technology-intensive industry was born in the United States, or that wizards of technology from Eli Whitney through Thomas Edison to Bill Gates have become American cultural icons. Technology is part of how Americans see themselves, to reach for it is instinctive. This works to the advantage of the American military.

That said, there will be new, radical technologies with great promise which will challenge the ability of the military to master and integrate. In particular, robotics, miniaturization, and nonlethality are likely to provide the keys to future success. But resistance to automated systems will probably be intense. It was the manned combat aircraft, carrier battlegroup, armored division, and Marine rifle squad that made the U.S. military the dominant one on earth. The natural tendency will be to reject revolutionary technology in favor of applique technology that augments the capability of existing combat systems in some marginal way. Eventually, the benefits of such an approach will peter out. The real issue will then be whether it is the U.S. military that first proves willing to jettison the old and adopt the new, or some other. History holds ominous warnings on this account. Seldom did the state that first began a revolution in military affairs end up mastering it. At one time, Britain and France were far ahead of Germany in aircraft and tank development. But Germany was more willing to innovate, in large part because of its military inferiority. The result was *blitzkrieg*. It is at least possible that the United States is blazing the way in the current revolution in military affairs, but will eventually be passed by another state or another nonstate actor driven to innovation by desperation and perceived weakness. In any case, the future certainly belongs to those armed forces willing and able to integrate new technology and squeeze the maximum effectiveness from it.

ANTICIPATING SECOND AND THIRD ORDER EFFECTS

Because strategy and armed conflict are so complex, any action has a multitude of second order effects (and third, fourth, and so on). Often, these become vital determinants of the future. When President Franklin Roosevelt authorized the Manhattan Project, for instance, who could have anticipated that this decision would eventually help to increase the strategic significance of rural guerrilla war (by raising the risks associated with traditional great power war), solidify the political and economic ties between the United States and its former enemies, Germany and Japan (by providing cheap security through extended deterrence), generate the "space race" as the United States and the Soviet Union competed for the strategic "high ground," and inspire a global peace movement? In all likelihood, strategic decisions made today, particularly by the United States, will have equally profound second and third order effects on 21st century armed conflict.

Some of these second order effects will be strategic and political. To take one example, by vigorously pursuing a revolution in military affairs designed to augment power projection and, perhaps, to lessen the need for allies, the United States may very well encourage the strengthening of regional security structures designed to minimize the need for American involvement or intervention. In part, this is because Americans consistently fail to understand how intimidating the combination of U.S. military power, U.S. economic power, and American popular culture can be. To many people around the world, the fact that the United States currently faces no pressing threat to vital national interests yet is willing to spend billions, perhaps trillions of dollars to undertake a revolutionary improvement in its armed forces is frightening. Washington, they conclude, must have plans to use the postmodern American military to impose its will on others. Even though U.S. policymakers are actually pursuing the revolution in military affairs as a way to retain the political utility of armed force and maintain the global political and economic gains made after the Cold War, the second order effects of this action are likely to prove adverse to American interests and objectives.

Many future innovations will bring equally profound and equally unexpected second and third order effects. The development of military robotics, biotechnology, and psychotechnology, in particular, may unleash a hurricane of unpredictable political, legal, and ethical problems. They may make armed force more precise and thus less horrific, or they may have the exact opposite effect and escalate the human costs to even higher levels. The growing interconnectedness among aggressor organizations may make the world a more dangerous place, or they may inspire greater interconnectedness among organizations dedicated to peace and security, thus making the world safer. The benevolence or malevolence of change is not preordained.

Ultimately, no one can fully predict the second order effects of innovations, much less third and fourth order effects. But this does not justify ignoring them. Any innovation, whether the development of new technology or the creation of new organizations and operational concepts, should be assessed for its broader political, social, cultural, ethical, and legal ramifications. It is probably unreasonable to expect militaries, even the most astute and farsighted ones like the U.S. military, to do this on its own. As interconnectedness links diverse things, shaping the future of armed conflict increasingly will become a shared task among militaries and other agencies and organizations. As the creator of first postmodern military, the United States must pave the way in this. Even this will not allow all second and third order effects to be *controlled*, but the more they can be anticipated, the better the decisions will be on questions of accepting or rejecting change and innovation. Successful militaries in the 21st century will thus be those which create a seamless web with nonmilitary organizations and agencies designed, in part, to anticipate second and third order effects.

PART IV: CONCLUSION AND RECOMMENDATIONS

CONCLUSION

Historians will see the last decade of the 20th century and the first decade or two of the 21st century as a turning point in the evolution of armed conflict. At this point we know fundamental change is underway but can only guess its ultimate outcome. Having assumed responsibility for encouraging and sustaining security around the world, the United States has a huge stake in this shift. To a large extent, the ability of the U.S. military to adapt to changes in the nature of armed conflict will determine whether the result is a more stable world or a more dangerous one. So far, the U.S. military has undertaken substantial efforts to understand and master the changes underway in the nature of armed conflict. But all these remain encumbered by the historic successes of the 1980s and 1990s. If the future wars which the U.S. military thinks about and plans for continue to look like reprises of the Gulf War or an updated version of a Warsaw Pact strike to the west, the American military may face 21st century war unprepared.

No nation has ever undertaken a full revolution in military affairs unless it is responding to perceived risk or recent disaster. The paralysis of victory is great and vested interests always powerful. If historical patterns hold, the U.S. military may not be able to make the leap into the future on its own. It often seems that the Pentagon's plans for the future, including systems acquisitions, are based on "bygone battles" [154]. Even the prestigious Defense Science Board has questioned whether Pentagon leaders are willing take the risks necessary to transform the military [155]. Ultimately, firm prodding may be necessary. This could come from one of two directions. One is direct and persistent intervention by its political masters. This might come from Congress. The Goldwater-Nichols Department of Defense Reorganization Act established a precedent for congressional intervention in the armed forces. Congress felt that after years of cajoling, the U.S. military was not taking jointness as seriously as it should. Legislation was used to change this. Outside intervention could also come from the President and Secretary of Defense if they were reform minded and willing to fight the inertia in the military and in the wider defense community. The second possible motive for revolutionary transformation, though, might be battlefield defeat. Just as the Battle of Jena led Prussia to serious military reform and defeat in World War I led Germany toward *blitzkrieg*, a bloody fiasco—if it did not cause an American withdrawal from global engagement— might fuel a revolutionary transformation within the military. If the nation is lucky, visionary leadership rather than American blood will inspire the necessary changes.

RECOMMENDATIONS

The key strategic challenges for the Army in the short- to mid-term (5 to 20 years) will be attaining greater strategic mobility, completing digitization, and becoming as effective at shaping the strategic environment as it is at responding to threats. The key strategic challenges for the mid- to long-term (15-30 years) will be:

Part IV: Conclusion and Recommendations

- developing and integrating robotics and miniaturized systems;
- stressing the full modularity of equipment, systems and organization;
- developing methods for the rapid transformation of doctrine, concepts, and organizations; and,
- developing greater psychological precision, including the full integration of nonlethal capabilities.

To prepare for this second wave of transformation, the Army should use its futures-oriented programs and intellectual resources, particularly the Army After Next Project and the War College, to explore the strategic implications of these challenges.

ENDNOTES

This study grew from a series of talks on the revolution in military affairs and future war given at locations ranging from Ouagadougou, Burkina Faso to the White House Conference Center. I owe a deep debt to those who invited me to speak and those who prodded me with questions or observations following the talks. I'd also like to thank Robert Bunker, James Wirtz, Edward Greisch, Douglas Johnson, Earl Tilford, Douglas Schnelle, and John Garofano for insightful comments on earlier drafts of this study. All flaws and shortcomings that remain are purely my own and do so despite the best efforts of this group of cutting-edge strategic thinkers.

[1] See, for instance, Secretary of Defense William S. Cohen, *Annual Report to the President and the Congress*, Washington, DC: The Pentagon, 1999, (henceforth *Annual Report 1999*), Chapter 10; *Transforming Defense: National Security in the 21st Century*, Report of the National Defense Panel, December 1997, p. iii; Keith Thomas, ed., *The Revolution in Military Affairs: Warfare in the Information Age*, Canberra: Australian Defence Studies Centre, 1997; Lawrence Freedman, "Britain and the Revolution in Military Affairs," *Defense Analysis*, Vol. 14, No.1, April 1998, pp. 55-66; Yves Boyer, *Une Révolution dans les Affaires Militaires?*, Paris: Fondation pour les

Etudes de Défense, 1998; Robbin F. Laird and Holger H. Mey, *The Revolution in Military Affairs: Allied Perspectives*, Washington, DC: National Defense University Institute for National Strategic Studies, 1999; and *The Future Security Environment*, Kingston, Ontario: Canadian Army Directorate of Land Strategic Concepts, 1999. The author has given presentations on the revolution in military affairs in Australia, Canada, Belgium, France, Singapore, Germany, Japan, and South Africa. This suggests the extent of interest in the topic.

[2] Alvin and Heidi Toffler, *War and Anti-War: Survival at the Dawn of the 21st Century*, Boston: Little, Brown and Company, 1993.

[3] *New World Coming: American Security in the 21st Century*, Report on the Emerging Global Security Environment for the First Quarter of the 21st Century, Arlington, VA: United States Commission on National Security/21st Century, September 15, 1999, p. 7.

[4] Bill Gates, "The Next Step for Technology Is Universal Access," *Forbes ASAP*, October 1, 1999, reprinted at *http://www.forbes.com /asap/99/1004/045.htm*.

[5] *Internet Relay Chat* was created by Jarkko Oikarinen in 1988. Since starting in Finland, it has been used in over 60 countries around the world. IRC is a multi-user chat system, where people meet in "channels" (also known as "rooms," these are virtual places, usually with a certain topic of conversation) to talk in groups, or privately. There is no restriction to the number of people that can participate in a given discussion, or the number of channels that can be formed on IRC. For more on the genesis of IRC see http://www.mirc.co.uk/help/jarkko.txt and jarkko2.txt.

[6] For a tiny sample, see http://www.cruzio.com/~blackops/; http://www.angelfire.com/ne/identityonline/; http://www.Stormfront.org/; http://www.christian-aryannations.com/; http://www.natvan.com;/ http://www.ufo.pair.com/dossier/abfuct-kidnap.html; http:// www. kingidentity.-com/doctrine.htm; http://www.wckkkk.com/people. html; http://www.neters.com/ web/wwf2.shtml; http://www. fortean times. com/artic/95/mind.html; http://www.- alienjigsaw.com/ index. html; http://www.thewinds.org/index.html; http://www2-.mo-net.

com/~mlindste/7momilit.html; http://www. melvig.org/; http://www.-usaor.net/users/ipm/; and http://ad2k. com/Army.html.
[7] *New World Coming*, p. 1.
[8] Thomas L. Friedman, *The Lexus and the Olive Tree*, New York: Farrar, Straus, Giroux, 1999, pp. xv-xvi.
[9] Robert O. Keohane and Joseph S. Nye, Jr., "Power and Interdependence in the Information Age," *Foreign Affairs*, Vol. 77, No. 5, September/October 1998, p. 93.
[10] Jessica T. Mathews, "Power Shift," *Foreign Affairs*, Vol. 76, No. 1, January-February 1997, p. 50.
[11] Martin van Creveld, *The Rise and Decline of the State*, Cambridge, U.K.: Cambridge University Press, 1999, p. vii.
[12] *New World Coming*, p. 27.
[13] See Mohamed Mahathir, *A New Deal For Asia,* Subang Jaya, Malaysia: Pelanduk Publications, 1999.
[14] 14 Hans Moravec, *Robot: Mere Machine to Transcendent Mind*, New York: Oxford University Press, 1999, p.1.
[15] For an interesting attempt to identify such states, see Ralph Peters, "Spotting the Losers: Seven Signs of Non-Competitive States," *Parameters*, Vol. 28, No. 1, Spring 1998, pp. 36-47. Like most of Peters' articles, this one is brilliant but hindered by hyperbole and sensationalism.
[16] See Timothy Egan, "Free Speech vs. Free Trade," *New York Times*, December 5, 1999.
[17] *New World Coming*, p. 6.
[18] Carl H. Builder and Brian Nichiporuk, *Information Technologies and the Future of Land Warfare*, Santa Monica, CA: RAND Corporation, 1995, p. 35.
[19] Wolfgang H. Reinicke, "The Other World Wide Web: Global Public Policy Networks," *Foreign Policy*, No. 117, Winter 1999-2000, pp. 44-56.
[20] Norman C. Davis, "An Information-Based Revolution in Military Affairs," in John Arquilla and David Ronfeldt, eds., *In Athena's*

Camp: Preparing for Conflict in the Information Age, Santa Monica, CA: RAND Corporation, 1997, p. 83.

[21] Douglas Farah, "New Drug Smugglers Hold Tech Advantage," *Washington Post*, November 15, 1999, p. 1.

[22] Tim Johnson, "Colombian Guerrillas Amass Air Force," *Miami Herald*, November 15, 1999.

[23] Friedman, *The Lexus and the Olive Tree*, p. 27. Friedman uses the Lexus—the Japanese luxury automobile—as a metaphor for what is modern and ingrained with cutting edge technology.

[24] Discussions with the author, Bonn, Germany, September 1999. Van Creveld also makes this point in *The Rise and Decline of the State*.

[25] For an astute assessment of the process of proliferation and its implications for the United States, see Robert W. Chandler, *The New Face of War: Weapons of Mass Destruction and the Revitalization of America's Transoceanic Military Strategy*, McLean, VA: AMCODA Press, 1998. Also useful are the various publications of the National Defense University's Center for Counterproliferation Research. See *http://www.ndu.edu/inss/ccp/cenresh.html*.

[26] Article 51 of the Charter does note that member states have the right of individual or collective self-defense "until the Security Council has taken measures necessary to maintain international peace and security."

[27] Edward N. Luttwak, "Toward Post-Heroic Warfare," *Foreign Affairs*, Vol. 74, No. 3, May-June 1995, pp. 109-122.

[28] Some research suggests that this perception is inaccurate. Peter D. Feaver and Christopher Gelpi contend that American leaders "have bought into a powerful myth, born during the Vietnam War and cemented during the ill-fated Somalia action of October 1993, that Americans are casualty-shy." ("How Many Deaths Are Acceptable? A Surprising Answer," *Washington Post*, November 7, 1999, p. B3.) See also Andrew P.N. Erdmann, "The U.S. Presumption of Quick, Costless Wars," *Orbis*, Vol. 43, No. 3, Summer 1999, pp. 363-382; Don M. Snider, John A. Nagl, and Tony Pfaff, *Army Professionalism, the Military Ethic, and Officership in the 21st Century*, Carlisle

Barracks, PA: U.S. Army War College Strategic Studies Institute, 1999, pp. 22-26; and James Burk, "Public Support for Peacekeeping in Lebanon and Somalia: Assessing the Casualties Hypothesis," *Political Science Quarterly*, Vol. 114, No. 1, Spring 1999, pp. 53-78. Whether the perception of casualty aversion is or is not accurate, though, it does shape current American military planning and long-term force and strategy development.

[29] Martin Libicki, "Rethinking War: The Mouse's New Roar?" *Foreign Policy*, No. 117, Winter 1999-2000, p. 41.

[30] See "African Center for Strategic Studies Begins 1 November in Dakar," *USIS Washington File*, October 22, 1999, reprinted at *http://www.eucom.mil/programs/acri/usis/99oct22.htm*.

[31] On ACSS, see "DOD Launches African Center for Strategic Studies," press release from the Office of the Assistant Secretary of Defense (Public Affairs), July 22, 1999, reprinted at http://www.defenselink.mil/news/Jul1999/b07221999_bt343-99.ht ml. For information on Military Professional Resources Inter-national (MPRI), see their web site at http://www.mpri.com/about/. Some analysts are concerned that the growth of organizations like MPRI represents an attempt to bypass congressional oversight of American defense policy. See Bruce D. Grant, "U.S. Military Expertise for Sale: Private Military Consultants as a Tool of Foreign Policy," *Essays 1998*, Washington, DC: National War College Institute for National Strategic Studies, 1998.

[32] See David Isenberg, "Soldiers of Fortune, Ltd.: A Profile of Today's Private Sector Corporate Mercenary Firms," Center for Defense Information Monograph, November 1997; idem., "The New Mercenaries," *Christian Science Monitor*, October 13, 1998, p. 13; David Shearer, "Outsourcing War," *Foreign Policy*, No. 112, Fall 1998, pp. 68-81; Khareen Pech, "Executive Outcomes—A Corporate Conquest," in Jakkie Cilliers and Peggy Mason, eds., *Peace, Profit or Plunder? The Privatisation of Security in War-Torn African Societies*, Halfway House, South Africa: Institute for Security Studies, 1999; and Herbert M. Howe, "Private Security Forces and

African Stability: The Case of Executive Outcomes," *Journal of Modern African Studies*, Vol. 36, No. 2, 1998, pp. 307-331.

[33] Robert Block, "African Supplier of Mercenaries Shuts, Says It Wants to Give Peace a Chance," *Wall Street Journal*, December 11, 1998, p. A13. For background on the legislation that led to Executive Outcome's demise, see Mark Malan and Jakkie Cilliers, "Mercenaries and Mischief: The Regulation of Foreign Military Assistance Bill," *Institute for Security Studies Papers*, No. 25, September 1997.

[34] Jimmy Seepe, "SA's Dogs of War Move to E Europe," *Johannesburg Sowetan*, February 3, 1999, p. 7, reprinted on the FBIS Internet site, February 3, 1999.

[35] Jakkie Cilliers and Richard Cornwell, "Mercenaries and the Privatisation of Security in Africa," *African Security Review*, Vol. 8, No. 2, 1999, and, pp. 31-42.

[36] See Sun Tzu, *The Art of War*, Samuel B. Griffith, trans., London: Oxford University Press, 1963; B. H. Liddell Hart, *Strategy: The Indirect Approach*, New York: Frederick Praeger, 1954; and Edward N. Luttwak, *Strategy: The Logic of War and Peace*, Cambridge, MA: Belknap, 1987.

[37] In addition to these tactical and operational characteristics of postmodern militaries, Charles C. Moskos and James Burk contend that such forces will (1) focus on subnational and nonmilitary issues; (2) be composed of small professional armies with reserves sharing missions; (3) face a skeptical or apathetic public; (4) have a much larger role for civilian employees; and, (5) have women and homosexuals integrated or accepted. See "The Postmodern Military," in James Burk, ed., *The Adaptive Military: Armed Forces in a Turbulent World*, 2nd edition, New Brunswick, NJ: Transaction, 1998.

[38] For instance, Cohen, *Annual Report 1999*, Chapter 1; and William S. Cohen, *Report of the Quadrennial Defense Review*, Washington, DC: Department of Defense, 1997, Section 2.

[39] *Report of the Quadrennial Defense Review*, Section 7.

[40] The concept of a "system of systems" originated with former Vice Chairman of the Joint Chiefs of Staff, Admiral William A. Owens. See, for instance, his article "The American Revolution in Military Affairs," *Joint Force Quarterly*, No. 10, Winter, 1995/1996, pp. 37-38. For an excellent assessment of the role of sensor grids in modern warfare, see Martin C. Libicki, *Illuminating Tomorrow's War*, Washington, DC: National Defense University Institute for National Strategic Studies, 1999.

[41] See the description in Ryan Henry and C. Edward Peartree, "Military Theory and Information Warfare," in Ryan Henry and C. Edward Peartree, eds., *The Information Revolution and International Security*, Washington, DC: Center for Strategic and International Studies, 1998.

[42] The Joint Staff is at work on a document entitled *Joint Vision 2015* which is to expand and refine the ideas of *Joint Vision 2010*. Early drafts suggest that this will be a powerful, even profound statement which will correct many of the shortcomings of the earlier document.

[43] *Joint Vision 2010*, Washington, DC: Chairman of the Joint Chiefs of Staff, n.d., pp. 14-15.

[44] Ibid., p. 11.

[45] Ibid., p. 24.

[46] Briefing by Admiral Harold W. Gehman Jr., Commander in Chief (CINC) of U.S. Atlantic Command, to the Joint Experimentation Futures Workshop, Breezy Point Naval Air Station, November 3, 1998. For background, see Mark A. Johnstone, Stephen A. Ferrando, and Robert W. Critchlow, "Joint Experimentation: A Necessity for Future War," *Joint Force Quarterly*, No. 20, Autumn/Winter 1998-1999, pp. 15-24. Information on the JFCOM J-9/ Joint Experimentation staff sections is available at http://www.acom.mil/jexp.nsf.

[47] Each of the services as well as the Defense Intelligence Agency, Central Intelligence Agency and other components of the government have crafted a vision or a series of visions of the future geostrategic environment. The Joint Experimentation Program at the

U.S. Joint Forces Command may help fuse the service visions. It remains to be seen whether the National Security Council can provide an even higher linkage.

[48] *Army Vision 2010*, Washington, DC: Headquarters, Department of the Army, 1996, p. 2.

[49] Ibid., p. 3.

[50] *Battle Labs: Maintaining the Edge*, Fort Monroe, VA: United States Army Training and Doctrine Command, 1994, pp. 6 and 32-33. The name "Louisiana Maneuvers" was an allusion to a series of massive training exercises held in the fall of 1941 as the U.S. Army began preparation for World War II. See James W. Dunn, *The 1941 Louisiana Maneuvers* at http://www.wood.army.mil/engrmag/pb 59512/review.htm. The U.S. Army's Louisiana Manuever Task Force was disbanded in 1996 as it gave way to the Force XXI process.

[51] See *Force XXI...America's Army of the 21st Century*, Washington, DC: Department of the Army, 1995.

[52] For an excellent explanation of the role that wargaming plays in the Army After Next process, see Robert B. Killebrew, "Learning From Wargames: A Status Report," *Parameters*, Vol. 28, No. 1, Spring 1998, pp. 122-135.

[53] *Speed and Knowledge*, the annual report on the Army after Next Project to the Chief of Staff of the Army, July 1997, p. 9-10.

[54] Richard K. Betts, "The New Threat of Mass Destruction," *Foreign Affairs*, Vol. 77, No. 1, January/February 1998, p. 27.

[55] *Speed and Knowledge*, p. 9.

[56] Ibid.

[57] *Second Annual Report of the Army after Next Project*, Fort Monroe, VA: Headquarters, United States Army Training and Doctrine Command, December 7, 1998, pp. 11-13.

[58] See, for instance, Joseph A. Engelbrecht, Jr., et al., *Alternative Futures for 2025: Security Planning to Avoid Surprise*, a research paper presented to Air Force 2025, April 1996; William B. Osborne, et al., *Information Operations: A New War-Fighting Capability*, a research paper presented to Air Force 2025, August 1996; Jeffrey E.

Theiret, et al., *Hit 'em Where It Hurts: Strategic Attack in 2025*, a research paper presented to Air Force 2025, August 1996; Bruce W. Carmichael, et al., *Strikestar 2025*, a research paper presented to Air Force 2025, August 1996; and Edward F. Murphy, *Information Operations: Wisdom Warfare For 2025*, a research paper presented to Air Force 2025, August 1996. These papers are available at http://www.au.af.mil/au/2025/.

[59] http://208.198.29.7/mcwl-hot/home/index.html

[60] Cohen, *Annual Report 1999*, Chapter 11.

[61] See *Forward . . . From the Sea: The Navy Operational Concept*, Washington, DC: Department of the Navy, 1997; and Department of the Navy 1999 Posture Statement: America's 21st Century Force.

[62] John B. Nathman, "A Revolution in Strike Warfare," *Sea Power*, October 1999, pp. 28-33.

[63] See, for instance, the web page of the Chief of Naval Operations Strategic Studies Group at http://www.nwc.navy.mil/ssg/.

[64] See David S. Alberts, John J. Garstka, and Frederick P. Stein, *Network Centric Warfare: Developing and Leveraging Information Superiority*, Washington, DC: Center for Advanced Concepts and Technology, 1999.

[65] Arthur K. Cebrowski, *Sea, Space, Cyberspace: Borderless Domains*, speech presented February 26, 1999, reprinted at http://www.nwc.navy.mil/pres/speeches/borderless.htm. See also Arthur K. Cebrowski and John J. Garstka, "Network-Centric Wmation Warfare: Its Origin and Future," *Proceedings of the U.S. Naval Institute*, Vol. 124, No. 1, January 1998, pp. 28-35.

[66] *Report of the Quadrennial Defense Review*, Section 7.

[67] *Defense Science Board 1996 Summer Study Task Force on Tactics and Technology for 21st Century Military Superiority*, Volume I: Final Report, Washington, DC: Office of the Secretary of Defense, 1996.

[68] Ibid., p. II-10.

[69] Harlan K. Ullman and James P. Wade, *Shock and Awe: Achieving Rapid Dominance*, Washington, DC: National Defense University Press, 1996.

[70] *Second Annual Report of the Army After Next Project*, pp. 5-6; and, Cohen, *Annual Report 1999*, Chapter 11.

[71] Joint Pub 3-13, *Joint Doctrine for Information Operations*, October 9, 1998, p. II-10.

[72] *Information Warfare: A Strategy for Peace...The Decisive Edge in War*, Washington, DC: The Joint Staff, n.d., p.2.

[73] See the web pages of the Joint Robotics Program (http://www.jointrobotics.com/); the Joint Non-Lethal Weapons. Program (http://iis.marcorsyscom.usmc.mil/jnlwd/); and the Air Force Information Warfare Center (http://www.afiwc.aia.af.mil/). See also General Richard Myers, Commander in Chief of U.S. Space Command, special briefing on the current activities of U.S. Space Command, January 5, 2000, reprinted at http://www.defenselink.mil/news/Jan2000/t01052000_t104myer.html.

[74] For elaboration, see Andrew Krepinevich, Executive Director of the Center for Strategic and Budgetary Assessments, *Emerging Threats, Revolutionary Capabilities, and Military Transformation*, testimony before the Senate Armed Services Committee, Subcommittee on Emerging Threats and Capabilities, March 5, 1999, reprinted at http://www.csbahome.com/Publications/ETACtestimony.htm.

[75] Russell F. Weigley, *The American Way of War: A History of United States Military Strategy and Policy*, Bloomington: Indiana University Press, 1973.

[76] *Transforming Defense*, p. 42.

[77] Betts, *The New Threat of Mass Destruction*, pp. 26-41.

[78] Ralph Peters, "Our Soldiers, Their Cities," *Parameters*, Vol. 26, No. 1, Spring 1996, p. 43.

[79] Thomas E. Ricks, "Urban Warfare: Where Innovation Hasn't Helped," *Wall Street Journal*, October 12, 1999, p. 10. See also William G. Rosenau, "Every Room Is a New Battle: The Lessons of

Modern Urban Warfare," *Studies in Conflict and Terrorism*, Vol. 20, No. 4, 1997, p. 385.

[80] Peter J. Skibitski, "Draft GAO Study Criticizes Pentagon Urban Warfare Capability," *Inside the Navy*, February 7, 2000, p. 2.

[81] Major General Robert H. Scales, Jr., *Future Warfare*, Carlisle Barracks, PA: U.S. Army War College, 1999, pp. 177-178. This passage is from an article entitled "The Indirect Approach: How U.S. Military Forces Can Avoid the Pitfalls of Future Urban Warfare," *Armed Forces Journal International*, Volume 136, No. 3, October 1998. In this essay, Major General Scales proposes an "indirect approach" to urban operations. This would entail establishing a loose cordon around a city occupied by enemy forces and isolating it from the outside world. Technology would then be used to suppress information flows within the besieged city. Strikes would be mounted against decisive points, largely for psychological effects as American forces demonstrate mastery. Eventually the will of those holding the city would erode and it would "collapse on itself."

[82] Daryl G. Press, "Urban Warfare: Options, Problems, and the Future," *Marine Corps Gazette*, Vol. 83, No. 4, April 1999, pp. 14-18.

[83] *Transforming Defense*, p. 15.

[84] John R. Groves Jr., "Operations in Urban Environments," *Military Review*, Vol. 78, No. 4, July-August 1998, pp. 31-40. For an analysis of specific doctrinal needs, see Russell W. Glenn, *". . . We Band of Brothers": The Call for Joint Urban Operations Doctrine*, Santa Monica, CA: RAND Arroyo Center, 1999, pp. 19-52.

[85] See Gary Anderson, "Urban Warfare: Russia Shows What Not To Do," *Wall Street Journal*, January 25, 2000.

[86] Daniel Verton, "Urban Warfare Tech May Alter Corps," *Federal Computer Week*, March 15, 1999.

[87] Joel Garreau, "Point Men for a Revolution: Can the Marines Survive a Shift from Hierarchies to Networks?" *Washington Post*, March 6, 1999, p. 1.

[88] Russell W. Glenn, *Marching Under Darkening Skies: The American Military and the Impending Urban Operations Threat*, Santa Monica,

CA: RAND Arroyo Center, 1998, p. viii. See also Russell W. Glenn, et al., *Denying the Widow-Maker: Summary of Proceedings of the RAND-DBBL Conference on Military Operations on Urbanized Terrain*, Santa Monica, CA: RAND Arroyo Center, 1998.

[89] For discussion of the strategic implications and operational concepts associated with nonlethal weapons, see Douglas C. Lovelace, Jr. and Steven Metz, *Nonlethality and American Landpower: Strategic Context and Operational Concepts*, Carlisle Barracks, PA: U.S. Army War College Strategic Studies Institute, 1998. See also Malcolm Dando, *A New Form of Warfare: The Rise of Non-Lethal Weapons*, London: Brassey's, 1996; and Nick Lewer and Steven Schofield, *Non-Lethal Weapons: A Fatal Attraction?*, London: Zed, 1997. Information on U.S. military efforts on nonlethal weapons is available on the web site of the Joint Non-Lethal Weapons Program at http://iis.marcorsyscom.usmc. mil/jnlwd/.

[90] *Robotics Workshop 2020*, sponsored by he U.S. Army Research Laboratory, February 25-27, 1997, Jet Propulsion Laboratory, Pasadena, CA. Workshop report developed by the Strategic Assessment Center of Science Applications International Corporation, McLean, VA, p. B-31.

[91] Ibid., p. B-32.

[92] Ibid., p. B-34.

[93] "University of California Researchers Mate Human Cells With Circuitry," story posted on the CNN web site, February 25, 2000, http://cnn.com/2000/HEALTH/02/25/bio.chip.ap/index.html.

[94] See Celia W. Dugger and Barry Bearak, "You've Got the Bomb. So Do I. Now I Dare You to Fight," *New York Times*, January 16, 2000 (electronic download).

[95] Quoted in Seth Mydans, "For Inept Burmese Rebel Band, Only Death Was Clear," *New York Times*, January 26, 2000 (electronic download).

[96] See, for instance, http://www.sccs.swarthmore.edu/~justin/fzln/about.html and http://www.utexas.edu/students/nave/. The latter lists 44 Zapatista web sites.

[97] David Ronfeldt, John Arquilla, Graham E. Fuller, and Melissa Fuller, *The Zapatista "Social Netwar" in Mexico*, Santa Monica, CA: RAND Corporation, 1998.

[98] This idea that the American military periodically purges its institutional memory of counterinsurgency and then has to re-learn it is developed in Steven Metz, *Counterinsurgency: Strategy and the Phoenix of American Capability*, Carlisle Barracks, PA: U.S. Army War College Strategic Studies Institute, 1995.

[99] L.R. Beam, "Leaderless Resistance," reprinted at http://www2.monet.com/~mlindste/ledrless.html.

[100] John Arquilla and David Ronfeldt, "Looking Ahead: Preparing for Information-Age Conflict," in Arquilla and Ronfeldt, eds., In *Athena's Camp*, pp. 461-462. Nichiporuk and Builder explore this same idea in *Information Technologies and the Future of Land Warfare*.

[101] Max G. Manwaring, ed., *Gray Area Phenomena: Confronting the New World Disorder*, Boulder, CO: Westview, 1993.

[102] For instance, Aimal Kansi was captured in Pakistan for a 1993 shooting outside CIA headquarters, and in 1990 Humberto Alvarez Machain was seized in Guadalajara, Mexico for his role in the torture of an American Drug Enforcement Agent. See Anthony J. Donegan, "The United States' Extraterritorial Abduction of Alien Fugitives: A Due Process Standard," *New England International and Comparative Law Annual*, Vol. 3, 1997, reprinted at http://www.nesl.edu/annual/vol3/alien.htm; Zahid F. Ebrahim, "Dodging the Law of Extradition," *Chowk Interactive Magazine*, reprinted at http://www.chowk.com/CivicCenter/zebrahim_mar1398.html; and United States v. Alvarez-Machain (91-712), 504 U.S. 655 (1992), reprinted at http://supct.law.cornell.edu/test/ hermes/ 91-712.ZO .html.

[103] Guilio Douhet, *The Command of the Air*, Washington, DC: Office of Air Force History, 1983. See also David MacIsaac, "Voices from the Central Blue: The Air Power Theorists," in Peter Paret, ed., *Makers of Modern Strategy From Machiavelli to the Nuclear Age*, Princeton,

NJ: Princeton University Press, 1986; and Bernard Brodie, *Strategy in the Missile Age*, Princeton, NJ: Princeton University Press, 1959, particularly pages 82-88.

[104] For instance, Robert Jervis, *The Meaning of the Nuclear Revolution: Statecraft and the Prospect of Armageddon*, Ithaca, NY: Cornell University Press, 1989, pp. 4-8.

[105] See the Commission's web site at http://www.pccip.gov/.

[106] John J. Stanton, "Rules of Cyber War Baffle U.S. Government Agencies," *National Defense*, February 2000, p. 29. On the government's program for dealing with cybercrime, see David Johnston, "U.S. Officials Lay Out Plan to Fight Computer Attacks," *New York Times*, February 17, 2000 (electronic download).

[107] Reported on the CNN Internet site, http://www.cnn.com/2000/TECH/computing/02/14/hacker.security/index.html.

[108] "Strengthening Cyber Security through Public-Private Partnership," fact sheet released by the White House Office of the Press Secretary, February 15, 2000, reprinted at http://www.whitehouse.gov/library/PressReleases.cgi?date=1 &briefing=1. For detail, see *Defending America's Cyberspace: National Plan for Information Systems Protection Version 1.0*, Washington, DC: The White House, 2000.

[109] See Winn Schwartau, "Ethical Conundra of Information Warfare," in Alan D. Campen, Douglas H. Dearth, and R. Thomas Goodden, eds., *Cyberwar: Security, Strategy, and Conflict in the Information Age*, Fairfax, VA: AFCEA International Press, 1996. On the legal arguments surrounding information warfare, see Walter Sharp, Sr., *Cyberspace and the Use of Force*, Falls Church, VA: Aegis Research Corporation, 1999.

[110] Lisa Hoffman, "U.S. Opened Cyber-War During Kosovo Fight," *Washington Times*, October 24, 1999, p. C1.

[111] Bradley Graham, "Military Grappling With Guidelines For Cyber Warfare," *Washington Post*, November 8, 1999, p. A1.

[112] General Richard Myers, Commander in Chief of U.S. Space Command, special briefing on current activities of the U.S. Space Command, the Pentagon, January 5, 2000, transcript at http:!!www.defenselink.mil!news!Jan2000!t01052000_t104myer.ht ml. See also Bill Gertz, "U.S. Set to Take Warfare On-Line," *Washington Times*, January 6, 2000, p. 3. General Myers has since become Vice Chairman of the Joint Chiefs of Staff.

[113] Although thinking in terms of low intensity conflict rather than high technology, Martin van Creveld made this same argument in *The Transformation of War*, New York, Free Press, 1991, pp. 198-201.

[114] Moravec, *Robot*, p. 25.

[115] Ray Kurzweil, *The Age of Spiritual Machines: When Computers Exceed Human Intelligence*, New York: Viking, 1999.

[116] *Robotics Workshop 2020*, pp. B-2 to B-3.

[117] *STAR 21: Strategic Technologies for the Army of the Twenty-First Century*, Technology Forecast Assessments, Washington, DC: National Academy Press, 1993, p. 148.

[118] Defense Advanced Research Projects Agency Distributed Robotics Program, overview reprinted at http:!!www.darpa.mil!MTO!D Robotics!index.html.

[119] Lee Gomes, "It's a Bird! It's a Spy Plane!—Pentagon Funds Research Into Robin-Sized Robots," *Wall Street Journal*, April 6, 1999, p. B1.

[120] H. Lee Buchanan, Deputy Director, Defense Advanced Research Projects Agency, testimony before the Subcommittee on Military Research and Development, Committee on National Security, United States House of Representatives, February 27, 1997.

[121] Defense Advanced Research Projects Agency, "MEMS Project Vision Statement," reprinted at http:!!www.darpa.mil!MTO. /MEMS/frameset.html?Overview.html. See also Albert P. Pisano, "MEMS 2003 and Beyond A DARPA Vision of the Future of MEMS," a presentation reprinted at http:// www.darpa.mil/MTO/ MEMS/2003 /index.html.

[122] *New World Coming*, p. 8.

[123] *Robotics Workshop 2020*, p. B-26.

[124] James Adams, *The Next World War: Computers Are the Weapons and the Front Line Is Everywhere*, New York: Simon and Schuster, 1998, p. 125.

[125] Martin C. Libicki, *The Mesh and the Net: Speculations on Armed Conflict in a Time of Free Silicon*, Washington, DC: National Defense University Institute for National Strategic Studies, 1994, p. 23.

[126] Ibid., p. 21.

[127] Ibid., p. 24.

[128] Ibid., p. 29.

[129] *New World Coming*, p. 20.

[130] *Robotics Workshop 2020*, p. A-8.

[131] Lonnie D. Henley, "The RMA After Next," *Parameters*, Vol. 29, No. 4, Winter 1999/2000, pp. 46-57.

[132] See Eric Drexler and Chris Peterson, *Unbounding the Future: The Nanotechnology Revolution*, New York: William Morrow, 1991, reprinted full text at http://www.foresight.org/UTF/Unbound_LBW/index.html. See also the nanotechnology web page of the Foresight Institute at http://www.foresight.org/NanoRev/index.html.

[133] David Voss, "Moses of the Nanoworld," *Technology Review*, March/April 1999, pp. 60-62.

[134] Colin S. Gray, *Modern Strategy*, Oxford, UK: Oxford University Press, 1999.

[135] Carl von Clausewitz, *On War*, Princeton, NJ: Princeton University Press, 1976, p. 89. For analysis of the meaning and misuse of Clausewitz's "trinity," see Edward J. Villacres and Christopher Bassford, "Reclaiming the Clausewitzian Trinity," *Parameters*, Vol. 25, No. 3, Autumn 1995, pp. 9-19.

[136] For analysis, see James Riggins and David E. Snodgrass, "Halt Phase Plus Strategic Preclusion: Joint Solution to a Joint Problem," *Parameters*, Vol. 29, No. 3, Autumn 1999, pp. 70-85.

[137] For instance, General Eric K. Shinseki, address to the Eisenhower Luncheon, 45th Annual Meeting of the Association of the United States Army, October 12, 1999, reprinted at http://www.hqda.army.mil/ocsa/991012.htm; *Transforming Defense: National Security in the 21st Century*, Report of the National Defense Panel, December 1997, p. 33; *Joint Vision 2010*, p. 13; etc.

[138] George and Meredith Friedman, *The Future of War: Power, Technology and American World Dominance in the 21st Century*, New York: Crown, 1996, p. x.

[139] U.S. Army Chief Scientist A. Michael Andrews, interviewed by Ron Laurenzo, *Defense Week*, November 29, 1999, p. 6.

[140] Sun Tzu, *The Art of War*, p. 66.

[141] Clausewitz, *On War*, p. 90 and ff.

[142] For example, John B. Alexander, *Future War: Non-Lethal Weapons in Twenty-First Century Warfare*, New York: St. Martin's, 1999.

[143] Luttwak, *Strategy*, pp. 62-65. For an astute analysis of how this tendency affects the post-Gulf War U.S. military, see Daniel P. Bolger, "The Ghosts of Omdurman," *Parameters*, Vol. 21, No. 3, Autumn 1991, pp. 28-39.

[144] Libicki, *Illuminating Tomorrow's War*, p. 1.

[145] For instance, Michael Dewar, *The Art of Deception in Warfare*, Devon, Great Britain: David and Charles, 1989.

[146] Andrew F. Krepinevich, "Cavalry to Computer: The Pattern of Military Revolutions," *The National Interest*, No. 37, Fall 1994, p. 30.

[147] Charles J. Dunlap, Jr., *Technology and the 21st Century Battlefield: Recomplicating Moral Life for the Statesman and the Soldier*, Carlisle Barracks, PA: U.S. Army War College Strategic Studies Institute, 1999, pp. 13-19.

[148] Libicki, *The Mesh and the Net*, pp. 50-69. See also Martin C. Libicki and James A. Hazlett, "Do We Need an Information Corps?" *Joint Force Quarterly*, No. 2, Autumn 1993, pp. 88-97. While the idea of an information corps has not yet received serious consideration within the U.S. Department of Defense, the Pentagon's Defense

Information Systems Agency has set up a special joint task force known as the Computer Defense Network to deal with defensive information warfare. (Bill Gertz, "Internet Warfare Concerns Admiral," *Washington Times*, November 18, 1999, p. 1).

[149] Bill Gertz, "China Plots Winning Role in Cyberspace," *Washington Times*, November 17, 1999, p. 1. See also "China: Army Publishes Book On Information Warfare", excerpts from report by Chinese army newspaper *Jiefangjun Bao*, December 7, 1999, reported by the BBC Monitoring Summary of World Broadcasts, reprinted at http://www.infowar.com/mil_c4i/00/mil_c4i_012400b_j.shtml.

[150] Robert J. Bunker, "Higher Dimensional Warfighting: Bond-Relationship Targeting and Cybershielding," a paper presented at *The Future of War*, the Ivan Bloch Commemorative Conference, St. Petersburg, Russia, February 24-27, 1999.

[151] Edward N. Luttwak, "From Vietnam to *Desert Fox*: Civil-Military Relations in Modern Democracies," *Survival*, Vol. 41, No. 1, Spring 1999, pp. 99-112.

[152] The idea of a "crisis" in American civil-military relations is explored in Don M. Snider and Miranda A. Carlton-Carew, eds., *U.S. Civil-Military Relations: In Crisis or Transition?* See also Douglas V. Johnson and Steven Metz, *American Civil-Military Relations: New Issues, Enduring Problems*, Carlisle Barracks: U.S. Army War College, Strategic Studies Institute, 1995; Michael C. Desch, *Civilian Control of the Military: The Changing Security Environment*, Baltimore: Johns Hopkins University Press, 1999; and Sam C. Sarkesian, "The U.S. Military Must Find Its Voice," *Orbis*, Vol. 42, No. 3, Summer 1998, pp. 423-437.

[153] Charles J. Dunlap, Jr., "The Origins of the Military Coup of 2010," *Parameters*, Vol. 22, No, 4, Winter 1992-93, pp. 2-20.

[154] Bradley Graham, "Pentagon's Wish List: Based on Bygone Battles?" *Washington Post*, August 25, 1999, p. A3.

[155] Colin Clark, "DSB: Pentagon Lacks Urgency to Effect Change," *Defense News*, October 11, 1999, p. 3.

ABOUT THE AUTHOR

Steven Metz has been Research Professor of National Security Affairs in the Strategic Studies Institute since 1993. Prior to that, he served on the faculty of the Air War College, the U.S. Army Command and General Staff College, and several universities. He has also served as an advisor to U.S. political organizations and campaigns, testified in the U.S. Senate and House of Representatives, and spoken on military and security issues around the world. He is author of more than seventy articles, essays, and chapters on such topics as nuclear war, insurgency, U.S. policy in the Third World, military strategy, South African security policy, and U.N. peace operations. Dr. Metz's current research deals with the role of the American Army in the 21st century, the African security environment, and the effect of the information revolution on military strategy. He holds a B.A. in Philosophy and a MA in International Studies from the University of South Carolina, and a PhD in Political Science from the Johns Hopkins University.

INDEX

#

20th century, 17, 18, 28, 33, 42, 45, 55, 81
21st century, viii, x, xi, xii, xiv, xvii, xix, xx, xxi, xxiv, 6, 9, 10, 11, 12, 16, 19, 21, 24, 33, 35, 40, 43, 44, 48, 49, 55, 57, 64, 68, 72, 73, 76, 77, 79, 80, 81, 105

A

Afghanistan, 6, 7, 36
agencies, 8, 11, 49, 50, 80
aggregation, 8
aggression, xi, 5, 11, 13, 16, 27, 32, 40, 41, 65
aggressors, xi, 16, 33
Air Force, xiii, 24, 27, 29, 33, 34, 55, 88, 92, 94, 98
American intervention, xi, 16, 34
American military power, xiii, 24
American security, vii, xxiv, 30, 86
antagonists, xiv, 10, 11, 40, 68
armed conflict, vii, ix, x, xi, xii, xiv, xv, xvii, xviii, xix, xx, xxi, xxiv, 1, 9, 10, 12, 13, 14, 16, 18, 19, 21, 27, 30, 39, 40, 48, 49, 50, 51, 52, 56, 57, 63, 64, 66, 68, 69, 71, 72, 73, 76, 77, 78, 80, 81, 100
armed forces, xv, 23, 31, 48, 49, 51, 53, 78, 79, 82
authoritarianism, x, 8
authorities, 49
autonomy, 5, 40, 61

B

ballistic missiles, 25, 34
battlefield, xii, xvi, xxi, 13, 21, 22, 26, 31, 37, 57, 58, 59, 60, 61, 71, 82, 101
biotechnology, xxi, 27, 79
bombs, xv, 51
bullets, xv, 51

C

cities, 17, 36, 37, 41, 44
citizens, x, 6, 8, 44, 50
civil rights, xix, xx, 50, 74, 75
Cold War, 11, 29, 41, 42, 45, 47, 79
collateral, 46, 52, 53
collateral damage, 46, 52, 53

combatants, xii, xiv, 18, 19, 37, 38, 40, 43, 50
communication, 3, 7, 29, 39, 66
computer, xv, 39, 45, 51, 53, 55
computer viruses, xv, 51
computing, 57, 98
conflict, vii, ix, xi, xii, xiv, xv, xvii, xviii, xix, xx, xxi, xxiv, 1, 6, 9, 10, 12, 13, 14, 16, 17, 18, 19, 21, 25, 27, 30, 36, 39, 40, 41, 42, 43, 47, 48, 49, 50, 51, 52, 56, 57, 63, 64, 66, 68, 69, 71, 72, 73, 76, 77, 78, 80, 81, 99
conflict resolution, 43
Congress, xxi, 30, 54, 82, 85
counterinsurgency, xiv, 43, 46, 68, 97
creativity, xiii, 8, 19, 27, 30, 32, 66, 73
criminal activity, 48
criminal investigations, 54
criminals, 50, 73
cruise missiles, 10, 29
culture, xviii, xx, 4, 5, 40, 41, 69, 75, 77, 79
cyberattacks, xv, 40, 52, 53, 54
cyberspace, 50, 73

D

democracy, 9, 12, 41
Department of Defense, x, xii, 13, 14, 21, 22, 29, 30, 31, 32, 82, 90, 102
deployments, 36
destruction, xi, xiii, 10, 14, 22, 25, 29, 34, 36, 47, 52, 55, 56, 64, 70
dispersion, xi, 14, 17, 22, 23, 44, 59, 65, 76
diversity, xi, 17, 18
dominance, 22, 29, 31, 32, 39
drug trafficking, 46, 50

E

economic consequences, 47
economic crisis, 5

economic power, 79
economic problem, 77
economics, xiv, xxiv, 4, 40
education, 14, 41, 67
elected leaders, 75, 76
enemies, vii, x, xiii, xv, xvi, xviii, xix, 9, 12, 13, 27, 29, 32, 34, 35, 36, 41, 46, 47, 48, 50, 52, 60, 62, 64, 66, 71, 74, 76, 79
environment, vii, xi, xv, xvi, xvii, xxii, xxiv, 1, 9, 14, 18, 22, 24, 31, 37, 38, 43, 50, 57, 59, 62, 66, 73, 76, 82, 92, 105
ethical issues, xvi, 61
ethical standards, xvii, 64, 74

F

Federal Bureau of Investigation, 50
federal government, 54
financial, xi, 4, 5, 13, 16, 18, 47, 53
financial community, 47
financial institutions, 5
financial markets, 4
financial resources, xi, 13
firearms, 67
force, x, xi, xii, xiii, xiv, xvii, xx, 7, 9, 10, 11, 12, 13, 14, 17, 19, 23, 24, 25, 26, 30, 31, 32, 34, 40, 41, 42, 46, 47, 49, 50, 51, 52, 55, 61, 64, 65, 69, 70, 72, 73, 75, 79, 89, 102

G

geological history, xi, 17
global economy, 5, 11
global security, vii, xi, xvi, 5, 14, 62, 86
globalization, 4, 5, 6, 12, 13, 41, 42, 48, 70
governments, x, 5, 8, 41, 46

H

history, xi, xxiii, 1, 3, 9, 14, 17, 18, 44, 51, 54, 67, 77
House of Representatives, 99, 105
human, xvii, xxiv, 1, 6, 7, 9, 39, 42, 57, 59, 61, 64, 67, 69, 77, 79
human activity, xvii, 64
human condition, 1
human rights, 42, 69

I

ideology, xiv, 3, 7, 40
industry, 18, 54, 59, 77
information revolution, vii, ix, x, xi, xxiv, 1, 3, 4, 6, 7, 9, 11, 14, 42, 45, 64, 71, 72, 91, 105
information technology, xii, xiii, xv, xvii, xviii, 1, 2, 3, 4, 8, 10, 19, 23, 26, 27, 30, 32, 36, 45, 48, 51, 53, 54, 59, 60, 61, 64, 66, 69, 71, 75
infrastructure, xv, 15, 30, 51, 52, 53, 54, 55, 74, 75
institutions, xix, 8, 42, 74
insurgency, 24, 43, 45, 46, 47, 105
insurgents, xiv, 9, 13, 41, 43, 44, 45, 46, 47, 48
integration, xii, xvi, xxii, 4, 6, 7, 19, 60, 75, 83
intelligence, xi, xiii, xiv, 10, 13, 16, 36, 39, 44, 45, 46, 48, 50, 59, 61, 67
intelligence services, xi, 16
interconnectedness, ix, x, xi, xxiv, 1, 7, 8, 11, 12, 13, 14, 42, 47, 48, 55, 65, 71, 77, 79, 80
international law, 11, 50
intervention, xi, xiv, xxi, 6, 14, 16, 30, 34, 42, 43, 49, 79, 82

L

law enforcement, xv, 49, 50, 74
laws, 50, 61
leadership, xiii, xxi, 8, 17, 24, 41, 45, 56, 75, 82
legal issues, 53
legal protection, 50
legislation, 90
logic bombs, xv, 51
logistics, 13, 23, 26, 30, 35, 39

M

machinery, 77
Marines, xiii, 28, 29, 38, 95
military, vii, viii, ix, x, xi, xii, xiii, xiv, xv, xvi, xvii, xviii, xix, xx, xxi, xxiii, xxiv, 1, 5, 8, 10, 11, 12, 13, 14, 15, 16, 18, 19, 21, 22, 23, 24, 25, 26, 27, 28, 29, 30, 31, 32, 33, 34, 37, 38, 39, 40, 41, 42, 43, 46, 49, 50, 51, 52, 53, 54, 55, 56, 57, 58, 59, 60, 61, 63, 64, 65, 66, 67, 68, 69, 71, 72, 73, 74, 75, 76, 77, 78, 79, 80, 81, 82, 85, 86, 89, 96, 97, 101, 102, 105
military affairs, vii, xvi, xvii, xxi, xxiii, xxiv, 1, 22, 29, 30, 31, 34, 37, 56, 61, 63, 64, 67, 71, 78, 79, 82, 85, 87, 91
military history, xi, 17, 51
military services, x, 13
militias, xiv, 13, 18, 40, 41, 43, 47, 49
missiles, xv, 10, 25, 29, 34, 36, 51
mission, 25, 29, 31, 39, 55, 57, 90
multidimensional, vii, xi, 23

N

nanotechnology, vii, 27, 28, 58, 60, 100
national interests, 79
national security, xv, 14, 24, 49, 50, 53, 75

natural resources, 48
nonlethal weapons, xiii, xiv, 33, 38, 46, 68, 70, 96
nuclear weapons, 40, 51

O

operations, xiii, xvi, xvii, xviii, 15, 17, 18, 19, 22, 23, 24, 26, 29, 32, 33, 34, 35, 37, 38, 39, 43, 45, 46, 52, 55, 64, 65, 66, 67, 75, 76, 95, 105
organized crime, xv, 47

P

paradigm shift, 28, 30
peace, 11, 13, 27, 79, 80, 88, 105
peacekeeping, 13
peacetime military, xi, 15
Pentagon, xxi, 30, 57, 82, 85, 95, 99, 102, 103
petroleum, 7, 27, 48, 65
police, 49, 50
policy, xi, 16, 21, 36, 40, 54, 89, 105
policy issues, 54
policymakers, xv, xxiii, 11, 12, 21, 53, 75
political leaders, 5, 30, 32, 55, 75
political opposition, 52
political organizations, x, 7, 105
political parties, 49
political power, 49
political problems, 50
political system, 11
politics, ix, xxiv, xxv, 4, 7
private sector, 61, 72
privatization, vii, xi, xviii, 14, 15, 16, 70
productive capacity, 52
proliferation, xvi, 10, 40, 47, 56, 59, 61, 70, 88
protection, xiii, xiv, 13, 22, 23, 29, 33, 46

psychotechnology, xvi, xx, xxi, 61, 69, 75, 79
public opinion, 6, 12
public policy, 8
public support, 36, 74

R

resolution, x, xiv, 13, 39, 43, 76
resources, xix, xxii, 8, 9, 18, 40, 66, 74, 75, 83
response, 13, 15, 23, 29, 46, 50, 52
risk, x, xxi, 3, 4, 8, 26, 38, 46, 47, 65, 66, 70, 74, 78, 82
risk aversion, 66
robotics, vii, xiii, xiv, xvi, xx, xxi, xxii, 28, 31, 33, 38, 39, 46, 57, 59, 60, 78, 79, 83, 94, 96, 99, 100

S

science, 18, 43, 60, 62
Secretary of Defense, xxi, 14, 21, 82, 85, 89, 93
security, vii, ix, xi, xv, xvi, xxi, xxiii, xxiv, 5, 9, 11, 14, 15, 16, 24, 26, 30, 35, 36, 43, 44, 46, 48, 49, 50, 54, 59, 62, 70, 71, 76, 79, 80, 81, 88, 98, 105
security forces, xv, 44, 46, 48, 49, 50
security services, 9
sensors, 22, 23, 30, 36, 39, 57, 59, 60, 61, 71
services, iv, x, xi, xii, xix, 2, 13, 16, 18, 23, 24, 27, 28, 29, 30, 31, 32, 38, 72, 74, 91
society, 41, 55, 56, 61, 73, 75
strategic partnerships, x, 7
strategic planning, 25, 48, 69
submarines, 25, 29
surveillance, 57

T

tactics, xvi, 17, 18, 19, 28, 33, 55
tanks, xviii, 25, 71
target, xii, xiii, 19, 22, 29, 39, 52, 56, 57, 61, 67
technological advancement, 3, 18
technological advances, 23
technological developments, 21
technological revolution, 60
technology, ix, xi, xii, xiii, xiv, xv, xvi, xvii, xviii, xx, xxiv, xxv, 1, 2, 3, 4, 7, 8, 9, 10, 13, 17, 19, 22, 23, 24, 25, 26, 27, 28, 29, 30, 32, 36, 38, 45, 46, 48, 51, 52, 53, 55, 56, 57, 58, 59, 60, 61, 62, 63, 64, 66, 68, 69, 70, 71, 73, 75, 77, 78, 80, 88, 99
tensions, xxiii, 47, 72, 74
territory, 25, 41, 48, 55, 65
terrorism, xi, xiii, 5, 13, 14, 17, 24, 36, 42, 44, 45, 50, 52, 56, 76, 95
terrorist attack, 36
terrorists, 13, 36, 41, 45
threats, xv, xviii, xxii, 37, 48, 49, 50, 69, 75, 82
three-dimensional space, 73
training, xvii, 14, 17, 38, 63, 64, 76, 92
Trojan horses, xv, 51

U

U.S. Air Force, xiii, 27, 55
U.S. Department of Defense, x, 13, 102
U.S. history, 48
U.S. military, vii, viii, ix, xi, xii, xiii, xiv, xviii, xix, xx, xxi, xxv, 14, 16, 19, 21, 22, 23, 27, 31, 32, 33, 37, 38, 40, 44, 46, 54, 55, 61, 68, 69, 73, 74, 75, 76, 77, 78, 79, 80, 81, 82, 89, 95, 96, 101, 102
U.S. policy, 79, 105
United Nations, 11, 43
United States, vii, xi, xiii, xiv, xv, xvii, xviii, xx, xxi, xxiii, 2, 4, 5, 7, 11, 12, 13, 14, 15, 16, 23, 25, 27, 32, 33, 34, 35, 40, 41, 43, 44, 45, 47, 48, 50, 52, 53, 54, 61, 64, 65, 68, 69, 70, 71, 73, 74, 75, 76, 77, 78, 79, 80, 81, 86, 88, 92, 94, 97, 99, 101

V

violence, x, 6, 9, 11, 40, 41, 47, 49, 70

W

war, xii, xiii, xiv, xv, xviii, xix, xx, xxiii, xxiv, 10, 11, 13, 15, 16, 17, 21, 22, 24, 27, 28, 29, 31, 33, 34, 36, 40, 41, 42, 43, 44, 47, 48, 49, 50, 51, 52, 53, 54, 55, 56, 58, 59, 63, 66, 68, 69, 70, 72, 74, 75, 76, 77, 78, 81, 85, 105
war crimes, 55
warfare, vii, xii, xiii, xv, xxiv, 10, 11, 15, 16, 17, 18, 21, 23, 24, 27, 28, 29, 32, 33, 36, 38, 39, 42, 44, 45, 46, 50, 51, 52, 53, 54, 55, 57, 59, 60, 61, 68, 70, 73, 74, 85, 87, 88, 91, 93, 94, 95, 96, 97, 98, 99, 101, 102
weapons, xi, xiii, xiv, 10, 14, 22, 23, 25, 26, 29, 30, 33, 34, 36, 37, 38, 40, 45, 46, 47, 51, 53, 56, 57, 60, 61, 64, 68, 70, 96
weapons of mass destruction, xi, xiii, 10, 14, 22, 25, 29, 34, 36, 47, 56, 64, 70, 88
web, x, 4, 8, 28, 35, 37, 42, 45, 80, 86, 89, 93, 94, 96, 97, 98, 100
worms, xv, 51